2nd Fully Revised and Enlarged Edition

Diagnostic Veterinary Parasitology

An Introduction

NIPA GENX ELECTRONIC RESOURCES & SOLUTIONS P. LTD.

New Delhi-110 034

About the Authors

Professor Dr Prayag Dutt Juyal born on January 5th 1955 at Pauri Garhwal Uttarakhand, did his BVSC & AH, MVSC (1979) from G.B. Pant University of Agricultural and Technology, Pantnagar and Ph.D (1987) from CCS, HAU, Hisar.

He served as Assistant Professor, Associate Professor and Professor in Veterinary Parasitology at R.A.U, Pusa (Bihar), G.B. Pant University of Agriculture and Technology, Pantnagar (Uttarakhand) and PAU/GADVASU, Ludhiana (Punjab). He was elevated to the position of Registrar, GADVASU, Ludhiana. He served GADVASU with retirement in 2015. Later in 2016, he become vice-chancellor of Nanaji Deshmukh Veterinary Science University, Jabalpur (Madhya Pradesh). He has more than 200 publications including research, popular and technical papers in national and international journal of repute. He has published 3 books and 5 practical manuals for PG and UG students respectively.

He is life member and fellow of many national level scientific societies. He was former vice-president and ex-general Secretary of Indian Association for Advancement of Veterinary Parasitology. He is recipient of Dr. BV Rao Gold Medal and Dr. Alwar Memorial Award for best research paper published. He is editorial board member of Journal of Animal Science Research and Journal of Laboratory Physicians AIIMS, New Delhi. He visited Thailand, U.K. Japan and China for presentation of his research findings. He was foliated with Biored Honorary Fellowship, Krishi Bhushan Award and Samgra life time achievement award.

Dr. Giridhari Das, born on November 15^{5th}, 1973 at Patrasayer, Bankura (West Bengal), did his B.V.Sc & A.H. from W.B.U.A.F.S. (Kolkata), and M.V.Sc and Ph.D (Veterinary Parasitology) form IVRI, Izatnagar, Bareilly. He has published more than 100 research and extension articles in reputed national and Internationals journals, five practical manuals for undergraduates students and nine practical manuals for postgraduate students. He is life member of many national scientific societies. He is recipient of Dr. V.S. Alwar Memorial Award for best research article published. 7Presently, he is serving as Director of Instruction and Controller of Examination and Professor and Head, Department of Veterinary Parasitology, Nanaji Deshmukh Veterinary Science University, Jabalpur.

2nd Fully Revised and Enlarged Edition

Diagnostic Veterinary Parasitology

An Introduction

Prayag Dutt Juyal MVSc, Ph.D, FIAAP, FNAVSc, FNADSI

Former Vice Chancellor

Nanaji Deshmukh Veterinary Sciences University, Jabalpur

Giridhari Das MVSc, Ph.D. (IVRI)

Professor and Head

Department of Veterinary Parasitology

Nanaji Deshmukh Veterinary Science University

Jabalpur, Madhya Pradesh

NIPA GENX ELECTRONIC RESOURCES & SOLUTIONS P. LTD.

New Delhi-110 034

NIPA GENX ELECTRONIC
RESOURCES & SOLUTIONS P. LTD.

101,103, Vikas Surya Plaza, CU Block
L.S.C. Market, Pitam Pura, New Delhi-110 034
Ph : +91 11 27341616, 27341717, 27341718
E-mail: newindiapublishingagency@gmail.com
www: www.nipabooks.com

For customer assistance, please contact
Phone: + 91-11-27 34 17 17
Fax: + 91-11-27 34 16 16

ISBN: 978-93-90175-00-0

Composed and Designed by NIPA.

Dedicated
to millions of animals
who suffered and laid their life
during investigations for the welfare of
mankind.................

Acknowledgement

Sincere thanks are due Dr N K Singh and Dr Harkirat Singh for their contribution in the earlier edition of this book. If any material contributed by them in earlier edition has been used in this edition also, we are indebted to them for that.

Authors

Preface to the Second Edition

Diagnosis is the integral part of control strategies against parasitic diseases in domestic animals. Identification of parasites by microscopy is the basis for both diagnostics and epidemiological assessment of parasite burden of a host. Veterinary clinicians usually face difficulty in proper diagnosis of clinical and subclinical cases of parasitosis in farm animals, pets and poultry because of lack of requisite information on commonly used diagnostic procedures at one place. This happens even though more than half of the cases reported in veterinary clinics and hospitals are of parasitic origin. Although various diagnostic techniques are amply available in textbooks and manuals, yet these procedures are scattered in different compilations that are mostly available in libraries, thus not easily accessible to field diagnosticians. Therefore, we attempted to compile easily adoptable diagnostic techniques for use of field veterinarians and students with the ultimate aim for substantial control of different types of parasitic infections and infestations in farm animals, pets and poultry. Additionally, this edition has chemotherapy and chemoprophylaxis against parasites.

Immunological and molecular techniques are increasingly being applied for the diagnosis of parasitic diseases, veterinary practitioners worldwide still rely primarily on the conventional faecal and blood smear examinations mainly on the concept of "seeing is believing". These techniques are still the gold standard and cost-effective means of diagnosing the parasitic infections in domestic animals. Our sincere thanks are due to all the faculty members and students of Department of Veterinary Parasitology for their facilitation in preparation of the book. Special thanks are due to (Late) Dr B S Gill, Ex-Dean, PAU, Ludhiana for his critical evaluation and invaluable inputs for improving the overall quality of the book.

Any suggestions from the readers for improvement of this book will be highly appreciated.

Authors

Preface to the First Edition

Diagnosis is an integral part of control strategies against parasitic diseases in domestic animals. Identification of parasites by microscopy is the basis for both diagnostics and epidemiological assessment of parasite burden of the host. Veterinary clinicians usually face difficulty in proper diagnosis of clinical and subclinical cases of parasitosis in farm animals, pets and poultry due to non availability of requisite information on commonly used diagnostic procedures at one place despite more than half of the cases reported in veterinary clinics and hospitals are of parasitic origin. Although various diagnostic techniques are amply available in textbooks and manuals, yet these procedures lie distributed in different compilations that are mostly available in libraries, thus not easily accessible to field diagnosticians. Therefore, an attempt has been made by the authors to compile easily adoptable diagnostic techniques for use of field veterinarians as well as students with the ultimate aim for substantial control of different types of parasitic infections and infestations in farm animals, pets and poultry.

Although immunological and molecular techniques are increasingly being applied for the diagnosis of parasitic diseases, veterinary practitioners worldwide still rely primarily on the conventional faecal and blood smear examinations based on the concept of "seeing is believing". These techniques are still the gold standard and cost-effective means of diagnosing the parasitic infections in domestic animals. Our sincere thanks are due to all the faculty members and postgraduate students of Department of Veterinary Parasitology for their facilitation in preparation of the book. Special thanks are due to (Late) Dr B S Gill, Ex-Dean, for his critical evaluation and invaluable inputs for improving the overall quality of the book.

Any suggestions from the readers for improvement of this book will be highly appreciated

Authors

Contents

Introduction

Amongst all types of infectious diseases of domestic animals, parasitic infections and infestations continue to be most widespread in all seasons particularly in tropical and subtropical regions of the world. It is practically impossible to observe under present managerial practices prevalent in India that any individual animal or flock may remain free of one or more parasites. In general, parasites are mostly insidious in nature and commonly lead to majority of subclinical cases; however, occasionally outbreaks do occur. Hence animal parasitism results in huge economic losses to the livestock owners engaged in this sector.

To control parasitic infections and/or infestations effective treatment can best be possible when correct and timely diagnosis is made. For this purpose various parasitological and serological techniques are applied besides recently developed molecular diagnostic tests. For application of all such techniques, the diagnosticians need to be familiar with salient diagnostic features of parasites. It must also be noted that nearly all organs and tissues of animal body are parasitized by one or more types/stages of parasites. Therefore, all body parts of living animals and their excretions/secretions *viz*. blood, lymph, spinal fluid, nasal and eye discharges, sputum, faeces and vomitus should be examined carefully for the presence of any developmental stage and/ or adult stage of parasites.

Sometimes small animals and birds severely suffer from parasitic infections and may result in moribund condition. Necropsy examination is suggestive of some of such cases particularly in case of outbreaks of coccidiosis and ascaridiosis in poultry; fasciolosis, immature paramphistomosis, severe gastro-intestinal nematodosis, piroplasmosis, trypanosomosis etc. in farm animals. Besides, post-mortem examination of animals died of acute parasitosis, should be done systematically (organ- wise) in addition to different tissues for search of parasitic stages to confirm the diagnosis. Necropsy is the most direct method to diagnose gastrointestinal parasitism during which some parasites like *Haemonchus*, *Bunostomum*, *Oesophagostomum*, *Trichuris* and *Chabertia* adults can be easily seen. However, infections with *Ostertagia*, *Trichostrongylus*, *Cooperia* and *Nematodirus* are difficult to see as these species are spread over meters of intestine, except by their movements in fluid ingesta. These smaller nematodes can be seen in gastrointestinal tract washings. For better understanding of localization of parasites in different organs and tissues, a table in the book has been appended for easy reference by diagnosticians.

This book would be of immense help for field veterinarians and students as it provides information on collection and preservation of parasites and their early developmental stages found inside or on the body of diseased animals for gross and microscopic examinations and for dispatch of such material to laboratory for diagnosis of causative parasite.

In addition to this, collection of excretions (preferably last of faeces untouched by ground/mud, urine and other contaminants), secretions (nasal discharge, sputum, vomitus), eyewash, vaginal wash, peripheral blood, lymph, skin scrapings (deep scrapings for mange lesions and sores), biopsy, impression smears from lymph glands, spleen, etc. are done for gross and microscopic examination. Physical examination of live animals is also essential, e.g. for enlarged lymph nodes, spleen and liver, holes in the skin, external ears, nodules in withers, external genitalia, mange lesions, attachment of ticks to help in arriving correct diagnosis made from other methods. The book also contains brief procedures of commonly used serological tests employed for early and prompt diagnosis in some parasitic infections in which no stage of parasite comes out of the body of the infected animal e.g. hydatidosis, cysticercosis, toxoplasmosis, haemoprotozoan infections and many others in farm animals, pets and poultry.

The details of various parasitological and molecular techniques for proper diagnosis of parasites of farm animals, pets and poultry are given in following pages along with illustrations wherever necessary.

1

Collection, Preservation and Dispatch of Faecal Samples

To identify parasitic infections, it is essential to examine excretory- secretory products (ESPs), *viz*., faeces, urine, sputum and nasal discharges as examination of such material is essential for clinical diagnosis. It is extremely rare to get an animal completely free from parasites and at the same time all animals may not show signs of disease. Most of the important helminths and coccidia inhabit intestinal tract. Therefore, clinical diagnosis of gut dwelling helminths and coccidian depends mostly on identification of ova/oocyst in faecal samples.

Collection of faecal sample

- Fresh faecal samples should be collected in a clean and dry wide mouth bottles (20-25 ml capacity). If that is not available, then faecal samples can also be collected in waxed-paper ice cream cups with sealing lid or even polythene bag with or without zip. There should not be any loss of moisture from the samples.
- Best way to collect the sample is manually directly from the rectum of the animals. For this, use disposable hand gloves, smear with vaseline or paraffin oil on the little finger (small animals) or entire hand (large animals).
- In pets, sometimes insertion of suppository or water enema may also be required to induce defaecation. It is better to ask the client actually witness the animal defecating to ensure the source of the samples and to note any straining, blood or mucous in the faeces.
- Faecal samples may have to be picked up from the ground in some cases. Make sure that the faeces should be fresh and free from extraneous matters or uncontaminated from urine, soil and dirt.
- The portion of faeces with mucus alone or with mucus and blood should be collected as these often show young or developing stages of the parasites in large numbers.
- If sample is old the protozoan parasites may disintegrate and nematode eggs will hatch within a few days.

- Any faecal samples that can't be examined within an hour after its collection should be refrigerated to slow or stop the parasite's development and reduce unpleasant odours.
- The minimum amount of faeces for examination is 2-5 g in case of pet animals. Whereas, it is 10 g for sheep and goats; and 50 g for large animals.
- Microscopic examination is more reliable when repeated.

Dispatch of faecal samples

Sometime field veterinarians may experience difficulties in conducting critical examination due to lack of facilities such as microscope or other equipments, or after having conducted the faecal examination themselves, may not be able to identify the parasites. For such cases dispatch of faecal samples to the nearest well-equipped laboratory for diagnosis is required.

For the dispatch of faecal sample the following procedure is adopted:

Put the faecal sample (approx. 10 - 20 grams) in a container of about 20 ml capacity and then fill the container with preservatives, (10% formalin for helminthic ova and 2.5% potassium dichromate solution for protozoan oocysts). Bottles/containers should be tightly stoppered (Fig.1) or sealed with lac or wax. For culture, fresh sample without preservative in a thermos with ice should be transported.

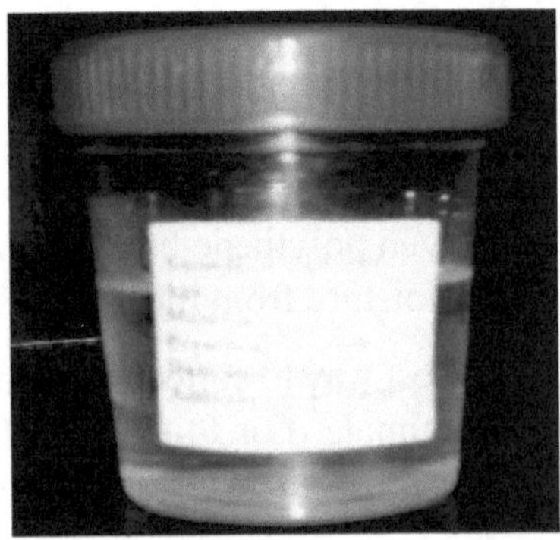

Fig. 1: Labeled vial containing 10% formalin

Labeling of the bottles or container with the following information is essential:

Proforma for Dispatching of Faecal Sample

Animal (species): ....

(Sex and breed): ....

Age: ....

Clinical signs (if any) ....

Material/sample: ....

Preservative used: ....

Date and time of collection and dispatch:

....

Examination required:

....

Name and Address of the
owner:...

Contact No

....

Treatment given, if
any:...

Signature with Stamp of Veterinary Officer Address: ....

....

....

Tel./Mobile No. ....

Email:..

Note: The sample should preferably be sent through a messenger or by any other quick mode of transport.

2

Techniques for Detection of Parasite Eggs/Oocysts in Faeces

Before proceeding to find out the nature of the egg content of the faecal sample, one should know what class of animal the faecal sample belongs to. It is easy to identify a faecal sample of herbivore from that of an omnivore or a carnivore. The dung of cattle, sheep and horse will contain lot of vegetable fibers. The colour of the dung under normal conditions will be dark greenish yellow with a peculiar aromatic smell. But the faeces of a dog or cat will be soft and pasty devoid of fibers and dark yellow in colour with an unpleasant smell.

Several procedures commonly used to examine faeces for parasites are described in this section. Before attempting to apply these procedures, a few points should be considered:

- Always handle faecal samples carefully. Some parasites, bacteria and viruses in animal faeces are a threat to human health. When examining the samples, the technician should always wear appropriate clothing and rubber or plastic gloves, clean the plateform thoroughly with a disinfectant soap after performing the tests.
- Always keep good records. A notebook should be kept in the laboratory area and every sample should be listed by the date, owner's name and the animal's name or number. Any observations about the appearance of the fecal sample, as well as any parasites found, should be recorded: otherwise it will be thought that the test has not yet been done. In recording negative findings, write "NPS" (No Parasites Seen).

Examination of faecal sample for helminthic infections

Gross Examination: It precedes the microscopic examination of faecal sample and includes collection of information as below:

a) **Consistency:** Condition of the faeces should be noted as soft, watery (diarrhoeic) or hard but it should be kept in mind that the description varies with the concerned animal species.

b) **Colour:** It is to report the unusual faecal colour, if any as mud coloured foul smelling faeces in calves indicate ascarid infection; tarry coloured faeces in dog is pinpointing of hookworm infection.

c) **Mucus:** Presence of mucus on surface of fresh faeces is indicative of intestinal parasitism particularly coccidiosis.

d) **Age of faeces:** It is always advocated to use fresh faecal sample as in aged samples the parasitic eggs may hatch and lead to misdiagnosis as the routine diagnostic tests are incompatible for detection of larvae.

e) **Gross parasites:** Faeces should be examined grossly for live nematode or dead worms, segments of tapeworms, e.g. gravid segments of *Moniezia* which resemble cooked rice grains, presence of cucumber seed-shaped segments in faeces of dog is indicative of *Dipylidium caninum*, immature amphistomes resembling pomegranate seeds (pink flukes) in diarrhoeic faeces of ruminants, larvae of certain arthropods (*Gasterophilus* spp.) in horses.

Microscopic Examination: Qualitative Methods

Direct smear method: A small quantity of faeces is placed on a clean, grease free glass microscopic slide. This is mixed with a drop of water and then spread out and covered with a cover-slip and examined directly under a microscope. This method though easy and simple to perform, is not sensitive and fully reliable. At least three direct smears should be examined from different parts of the faecal sample before declaring the sample negative.

Concentration Method

Principle: To concentrate all the eggs in a given amount of faeces into as small quantity of the emulsion as possible and then examining a drop of it. There are two primary types of concentration methods used in veterinary practices: floatation and sedimentation. Sedimentation is mainly used to detect eggs or cysts that have too high a specific gravity to float or that would be severely distorted by floatation solution. 0.5% glycerin in water can be used for emulsifying the faeces instead of plain water. It can be used for roundworm and tapeworm eggs, but there is usually too much faecal debris hiding the eggs to make it worthwhile. For that reason, this procedure is not used routinely and has its greatest use in suspected fluke infections.

Sedimentation

Simple sedimentation method

- Put 1-2 g of faecal material in mortar
- Add little quantity of water and triturate with pestle
- Add some more water, strain it through a sieve
- Transfer the mixture to a beaker or centrifuge tube (Fig.2)
- Keep it undisturbed for 10-20 minutes
- Discard the supernatant
- Transfer a drop of sediment to a glass slide and place a cover- slip
- Examine under low power (10X) of microscope in the manner shown in Fig. 3 for thorough examination of area under cover-slip.

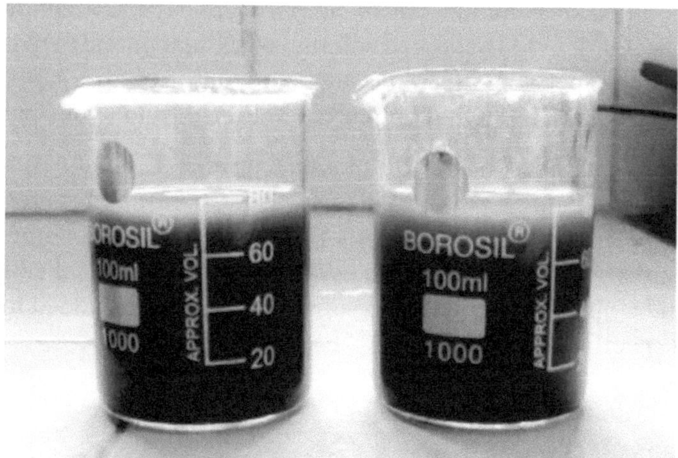

Fig. 2: Sedimentation in glass beakers

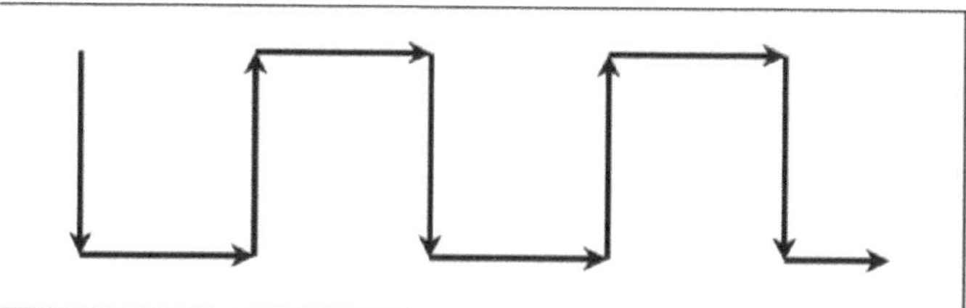

Fig. 3: Movement pattern of microscopic field

Sedimentation with centrifugation

- Place 1-2 g of faecal material in a mortar and then thoroughly mix it by adding little amount of water
- Strain through a sieve and transfer it to a centrifuge tube
- Make the volume to 15 ml with water and centrifuge at 1500 rpm for 5-10 minutes
- After centrifugation, discard the supernatant and repeat the process till the supernatant becomes clear
- Finally, transfer a drop of sediment to the slide, put cover-slip and examine under the microscope

Floatation

Principle: The principle is to use an emulsifying fluid of a greater specific gravity than that of the contained eggs. The common fluids in use are (1) Saturated solution of common salt of Sp. Gr. 1.18-1.19 (2) Saturated sugar solution of Sp. Gr. 1.25 (3) Saturated solution of sodium nitrate (4) Magnesium sulphate solution Sp. Gr 1.32 (5) Zinc sulphate solution of Sp. Gr. 1.18 (32.5% solution) and (6) A.E.X. (HCI of Sp. Gr. 1.03+Ether+Xylol equal parts).

Most parasite eggs have a specific gravity between 1.1 and 1.2, while tap water is only slightly higher than 1.000. Therefore, parasite eggs are too heavy to float in tap water. To make the eggs float, a liquid with a higher specific gravity than that of the eggs must be used. Such liquids are called floatation solutions and consist of concentrated sugar or various salts added to water to increase its specific gravity.

Simple floatation method

- Put 1-2 g of faeces in mortar
- Add little quantity of saturated salt solution
- Triturate with pestle
- Strain the mixture through a sieve
- Transfer to a glass/plastic vial
- Fill the vial up to the top with saturated salt solution so that a convex meniscus is formed
- Put a cover-slip over the top of the vial (Fig.4)
- Keep it undisturbed for 10-15 minutes

- Pick up the cover-slip gently, place it over glass slide and then examine it under a microscope

Fig. 4: Floatation technique using vials

Floatation with centrifugation

- About 1-2 g of faeces is transferred to a mortar
- This is moistened with small quantity of water and then thoroughly mixed with pestle
- The mixture is strained and transferred to centrifuge tube
- To this floatation solution (saturated salt solution or 33% Zinc sulphate solution) is added up to the brim of the tube
- The tube is then centrifuged at 1500 rpm for 3 to 5 minutes
- A drop of the sample from the top is then transferred to a slide and examined under the microscope using a cover-slip (cover- slip can also be used while centrifuging)
- Place round cover-slip on the brim of the centrifuge tube leaving a very small air bubble between cover-slip and fluid filled up to the brim. After centrifugation, pick up the cover-slip from its edges with a finger and thumb and immediately put on a clean slide. If positive, eggs/ova/cysts/oocysts will be seen in large numbers

Identification of Helminthes eggs and its Characteristics

***Fasciola* spp.**

Host: Cattle, Buffalo, Sheep, Goat, Pig and Elephant

Eggs are oval and have a thin shell, yellow tinged, operculated (indistinct)

Germinal cell evenly distributed, hexagonal

Measure 150-170 x 65-90 μm

Amphistomes

Host: Cattle, Buffalo, Sheep, Goat, Pig and Elephant

Eggs are thin shelled

Colourless, operculated (distinct)

Germinal cells are coarsely distributed or clumped

Comparatively smaller than *Fasciola* spp.

Measure 110-150 X 52-70 µm

Schistosoma nasalis

Host: Cattle, Buffalo

Characteristic boomerang shape

Transparent with thin shell

Spine present at one end

Miracidium present within the egg

Measure 330-580 X 50-80µm

Schistosoma indicum

Host: Cattle, Buffalo, Sheep, Goat Camel and Horse

Oval in shape with a terminal spine

Well developed miracidium in the egg

Schistosoma spindale

Host: Cattle, Buffalo

Spindle or napoleon hat shaped with one side flat

Thin shelled large egg

Spine present at one end

Schistosoma incognitum

Host: Pig and Dog

Oval in shape with a subterminal spine

Well developed miracidium in the egg

***Moniezia* spp.**

Host: Cattle, Goat and Sheep

Egg shell capsule is roughly square with rounded corners or somewhat triangular in shape. (*M. expansa*- triangular and *M. benedeni*-square)

Embryophore is pear shaped and is called pyriform apparatus and has a cap

Contains a hexacanth embryo

Size :60-80 μm in diameter

***Taenia* spp.**

Host: Cat, Dog, Man

Egg shell capsule absent

Small egg

Embryophore thick, round shape, radially striated

Hexacanth embryo present

Size : 35-80 μm in diameter

***Hymenolepis* spp.**

Host: Rats, Birds, Man

The egg shell is thin enclosing the hexacanth embryo

Wide space present between egg shell and embryophore

Comparatively larger than eggs of *Taenia* species

Size : 60-80 μm in diameter

Dipylidium caninum

Host: Dogs and Cats

Large oval capsules containing 4 to 20 eggs which have thick but unstriated embryophore

Size : 25-40 μm in diameter

Diphyllobothrium latum

Host: Dogs, Cats and Man

Small oval

Egg with thin shell (similar to trematode egg)

Ends appear pointed than round

Operculum present

Germinal mass undifferentiated and uniformly distributed

***Ascaris* spp.**

Host: Pig and Man

Sub-globular shape

Thick shell which is coarsely pitted and mammiliated

The unsegmented yolk appear as a compact mass in the centre

Size: 50-80 µm in diameter

Toxocara vitulorum

Host: Cattle and Buffalo

Round or sub-globular in shape and the egg shell is finely pitted

Size : 75-95µm x 60-75µm

Toxocara canis

Host: Dog and Cat

Sub-globular in shape and the egg shell is finely pitted

Size: 70-95 µm

Toxascaris leonina

Host: Dog and Cat

Large round

Egg shell which is smooth

Size : 75-85µm x 60-75µm

***Strongyloides* spp.**

Host:Cattle, Sheep, Goat, Horse, Dog, Bird

The comparatively small size and oval shape

The thin shell with both poles blunt

Well developed embryo or larva inside

Strongyle spp.

Host: All domestic animals

Eggs of nematodes belonging to Superfamilies- on tissue with Strongyloidea, Trichostrongyloidea

Oval shape, size varies with species (small, medium, large eggs)

Thin egg shell

Segmented yolk

Size: 60-70X 36-45 µm

Hook worm

Host: Dog, Cat, Cattle, Sheep, Goat and Man

Eggs of nematodes belonging to Superfamilies Ancylostomatoidea

Eggs are oval shape with blunt end and parallel side

Segmented yolk in 8-16 cells stage

Size: 60-70X 36-45 µm

***Trichuris* spp.**

Host: Cattle, Sheep, Goat, Dog, Man and Pig

Barrel shaped with thick shell

Dark brown coloured

Transparent plugs present at either poles

Yolk is unsegmented

Size : 30 -37 X 11-15 µm

Interpretation of qualitative examination of eggs (under low power field microscope)

- 1egg per field : Occasional
- 1-2 eggs per field : +
- 2-4 eggs per field : ++
- 4-6 eggs per field : +++
- >6 eggs per field : ++++
- No ova / cyst seen : Negative

Quantitative methods

Quantitative methods of the examination are employed to measure the magnitude of infection in terms of eggs per gram (EPG), oocyst per gram (OPG) in the faecal sample.

Stoll's egg counting method

- Take 3 g of faeces and transfer into a graduated flask tube
- Fill the tube flask upto 45 ml mark with N/10 Sodium hydroxide solution and add 10-12 glass beads
- Shake it vigorously to make a homogenous suspension of the faecal matter.
- Place 0.15 ml suspension on a slide cover with cover-slip and count all the eggs present
- Multiply the total number of eggs obtained by 100 to determine the EPG

McMaster's egg counting technique

- Take 3 g of faecal material and soak in 42 ml of saturated salt solution
- Triturate in a pestle and mortar and strain it
- Withdraw sample by means of a wide mouthed pipette and charge the McMaster's counting chamber (Fig.5)
- The number of eggs within each ruled area multiplied by 100 represents the EPG of the sample

Interpretation of Eggs per Gram Counts

EPG			Interpretation
Gastrointestinal nematodes			
Horses:	500	:	Mild
	800-1000	:	Moderate
	1500-2000	:	Severe
Cattle:	300-600	:	Severe
Lambs:	1000	:	Moderate
	2000-6000	:	Severe
Fasciola gigantica			
Cattle:	100-200	:	Severe
Sheep:	300-600	:	Severe

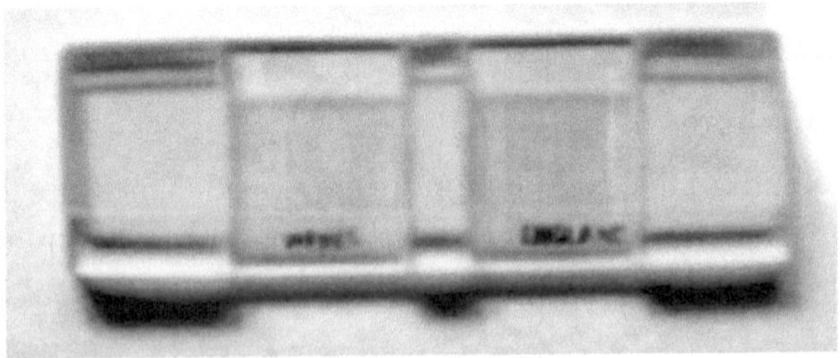

Fig. 5: McMaster Chamber

Faecal examination for enteric protozoan oocysts

Protozoans can be classified on the basis of their location as enteric, blood and tissue protozoan. Enteric protozoan generally encountered in intestine are amoeba, intestinal flagellates e.g. *Entamoeba, Giardia, Isospora, Balantidium* and *Histomonas.* Trophozoite stages of the protozoan are generally found in the diarrhoeic stool i.e. acute phase of disease. It is highly fragile and remains intact for only 15-20 minutes, so the faecal examination should be done immediately after collection of the faecal material. Cystic stages of the protozoa are found in well- formed stool i.e. in chronic phase of the disease. The protozoan cysts/ oocysts in the faecal sample are generally examined by simple floatation technique as described above with the exception of oocysts of *Eimeria leukarti* in horses which are detected in sedimentation technique as they are heavier. The faecal samples for protozoan infections may be preserved in polyvinyl alcohol to restore the morphology of trophozoites. Also wet faecal smears may be examined at the earliest to see the movements of trophozoites.

Iodine staining

It is done to differentiate between protozoan parasites and pseudo- parasites. Wet faecal smear may be stained with Lugol's iodine by mixing one drop of Lugol's iodine with one drop of faecal sample. By this staining the internal structures of the parasites are easily visualized thus differentiating them from pseudo-parasites.

Formol ether centrifugation method

It is done particularly in those parasitic infections in which a large amount of fat is passed out in faeces, such as amoebiosis and giardiosis.

Procedure

- Prepare a faecal suspension by mixing 1 part of faeces with 10-12 parts of saline
- Strain the suspension through the sieve and put the suspension in centrifuge tube upto mark of 15 ml and centrifuge at 2000 rpm for 2 minutes
- Discard the supernatant and add fresh saline solution to the sediment and repeat the centrifugation step twice
- Add 10 ml of 10% formalin to the sediment and leave it for few minutes
- Add 5 ml of ether and resuspend the sediment by vigorous shaking of the tube with gloved finger for 20-30 seconds and centrifuge again

- Decant the supernatant completely and break the white coloured formalin-ether junction using Pasteur pipette
- Transfer sufficient amount of sediment to slide, put cover-slip and examine under microscope

Identification of enteric protozoan cysts and its characteristics

Giardia lamblia

Host : **Man**

Location : Upper part of small intestine (Duodenum)

Disease : **'Travellers' diarrhoea**

Trophozoite

- Pyriform shaped body with bilaterally symmetrical
- Anterior end broadly rounded
- A large sucking disc on ventral surface
- Dorsal surface convex and ventrally concave
- Two nuclei, two axostyles and 8 flagella (3 pairs anterior and one pair posterior)
- A pair of dark stained median bodies or the axostyle

Cyst

- More or less pear shaped, oval or ellipsoidal
- Four nuclei concentrated at the pole
- A number of fibrillar remnants of the trophozoite organelles

Entamoeba histolytica

Host : **Man, Monkey, Dog, Cat, Pig, Rat, etc.**

Location : Ascending colon and vermiform appendix (Large intestine)

Disease : **Amoebic dysentery, amoebic granuloma, amoebic liver abscess**

Trophozoite

- Hyaline ectoplasm and finely granular endoplasm
- Nucleus has small central karyosome

- Long or short or blunt finger like pseudopodia
- Fine nuclear chromatin at the periphery of the nucleus

Cyst

- Round shape or ovoid (5-20 μm in diameter)
- Four distinct nuclei in the centre of the cyst
- Chromatoid bodies appear as refractile rods with rounded ends

Entamoeba coli

Host : Man

Cyst

- Round shape or ovoid (10-30 μm in diameter)
- Eight distinct nuclei
- Chromatoid bodies with splinter like ends

Balantidium coli

Host : Man, Pig and Monkeys

Location : Large intestine

Disease : Balantidiosis

Trophozoite

- Vegetative forms are oval shape and 50-60 μm in diameter, presence of cilia all around in longitudinal rows
- Peristome at the narrow anterior end
- Saucer or kidney shaped macronucleus and a small spherical micronucleus which lies in the notch of the macronucleus
- Presence of usually two contractile vacuoles and food vacuoles with particles, RBC's etc.

Cyst

- Ovoid to spherical in shape, 40-60 μm in diameter, faintly yellowish green in colour
- Organisms can be seen within the cyst wall by the presence of macronucleus and minute cilia

Histomonas meleagridis

Host : **Turkeys, Chicken and Guinea fowl**

Location : Caeca and liver

Disease : **Black-head, Infectious entero-hepatitis, Histomonosis**

Morphology

- Organisms are highly pleomorphic
- There are four stages: Invasive, Vegetative, Resistant and Flagellated stage
- In flagellated stage body is amoeboid
- Cytoplasm is composed of a clear outer ectoplasm and a coarsely granular endoplasm
- Nucleus is often vesicular, with a single dense karyosome
- Single flagellum is present

Procedure for acid fast/Ziehl Neelsen staining for cryptosporidial oocysts

- Make a translucent faecal smear with a cotton swab and air dry it by moving in the air
- Fixed the smear with methanol for 30 sec to 2 minute
- Pass the dried faecal smear through a flame for fixation
- Stain the fixed smear with Ziehl Neelsen carbol fuchsin solution for 40 minutes and rinse with tap water
- Rinse the stained smear for a few seconds with acid alcohol (3% HCl in 70% ethanol) that act as decolorizer and rinse again with tap water
- Counter stain the smear with brilliant green for 5 minutes and rinse with tap water
- Air dry the slide and examine under oil immersion lens of the microscope

Interpretation

Cryptosporidium oocysts will appear as bright red granules on a blue green background under oil immersion lens of microscope.

Morphology of *Cryptosporidium* oocyst

- 5-7.4μm x 4.5-6.5μm with colorless bilayered oocyst wall having an ultrastructurally faint suture like line
- Each sporulated oocyst contains four sporozoites laying parallel to one another
- After acid-fast staining (Ziehl-Neelsen) oocyst appeared as thick walled red granules

Sporulation of coccidian oocysts

Examination of the sporulated oocysts of coccidian parasites is done for differential diagnosis of different coccidian oocysts as they contain variable number of sporocysts and sporozoites which are species specific. Sporulated oocysts of *Eimeria* spp. have four sporocysts each containing two sporozoites whereas, *Isospora* spp. have two sporocysts each containing four sporozoites. Further, the sporulation time of different species within the same genera is different (from few hours to few days) that helps in identification of particular species within the genera. However, *Sarcocystis* oocysts are fully sporulated before being passed out in the faeces of definitive host so require no sporulation. This is the reason why they have high transmission potential. Sporulated oocysts (consisting of two sporocysts, each containing four sporozoites) may be broken and single sporocyst with four sporozoites is mostly seen.

Procedure

- Prepare 2.5% solution of Potassium dichromate ($K_2Cr_2O_7$)
- Mix a few grams of faeces with adequate volume of 2.5% Potassium dichromate solution (Potassium dichromate has nascent oxygen which act as oxidizing agent and also as preservative)
- Spread the mixture in thin layers in Petri dish such that the fluid layer should be 2-4 mm in thickness
- Keep the mixture in Petri dish at room temperature (27°C) and examine after 24 h intervals for 5-7 days which allows complete sporulation. Thorough blowing of culture medium in the dish is necessary for introduction of air or oxygen
- Examine a drop of the material under a microscope
- Majority of the oocysts will sporulate after 48 h. The material may also be transferred to a bottle and stored in a refrigerator at 4°C for further examination

Identification of coccidian oocysts and its characteristics

***Eimeria* spp.**

Host : Birds, Cattle, Sheep, Goat and other animals

Unsporulated Oocyst

- Oval or round in shape, with double contoured wall
- It contains germinal mass (zygote or sporont) either covering the whole inner space or contracted in the centre and in varying stages of development

Sporulated Oocyst (infective stage)

- Oval in shape
- Presence of 4 elongate ovoid sporocysts with one end pointed, each containing 2 comma shaped sporozoites
- Micropyle when present is at the anterior end. There may be a polar/ micropylar cap, stiedae body at the more pointed end of sporocyst, oocystic residual body, a polar granule and a secondary or sporocystic residual body inside each segment

***Isospora* spp.**

Host : Dog, Cat and Pig

Unsporulated oocyst

- Oval or round in shape
- Immature oocyst contains a germinal mass in different stages of development

Sporulated Oocyst

- Mature oocyst contain two sporocysts each sporocyst contains four sporozoites

Preparation of faecal cultures and recovery of larvae by Baermann's technique

Sometimes a specific diagnosis could not be arrived at from the eggs during faecal sample examination as in case of strongyle type eggs. Then it becomes necessary to culture the eggs for the harvesting of larvae in the laboratory.

There are basically 4 methods of faecal culture and most commonly used given below

1. Petridish method
2. Jar method
3. Harada Mori Test Tube Culture Method
4. Preparation of faecal cultures and harvesting or recovery of non-migratory larvae by Baermann's technique

Faecal Culture

- Positive faecal sample is broken up and mixed with charcoal and soil in the ratio of 3:2:1 in the Petri dishes
- Correct consistency should be like that of normal sheep faeces. If faeces are too dry then moisten them with water and if too wet then sterilized and dried faeces are added and a good grade of animal charcoal can be used thereafter
- Keep the sample at temperature of 26-27°C for 6-7 days to harvest the maximum number of infective third stage larvae (L_3)
- Regular sprinkling of water and stirring the faecal culture is required to avoid the growth of fungus and drying
- Recover the larvae from faecal cultures by placing the petri dish in a dull light or using Baermann's apparatus

Baermann's technique

This method was devised for the separation of larvae from soil, vegetation, grass blades and faecal culture. A glass funnel, 20 cm wide at the top and closed at the bottom with a clamped piece of rubber tubing is fixed in a stand and a circular piece of wire gauge about 9 cm in diameter (or tea sieve) is placed in the funnel and on it a large piece of linen/muslin cloth is placed (Fig. 6). The incubated faeces are spread 2 cm deep on the linen and water at 40°C is poured down the side of the funnel until it covers the faeces. The apparatus is left for 8- 12 h in dark until the larvae have migrated from the faeces into the water and sediment into the neck of the funnel so that will be found in first few millilitres of water drawn off the bottom. If few or no larvae are seen, a considerable amount of water should be withdrawn, centrifuged and the sediment should be examined.

Fig. 6: Baermann's Apparatus

Identification of nematode (strongyle) larvae

A drop of preserved sediment containing larvae was placed on a glass slide, mixed with a drop of Lugol's iodine or aqueous Safranin and then examined under dry magnification of the compound microscope after applying a cover slip over the preparation. Identification of the strongyle larvae was done with the help of keys and plates provided by Ministry of Agriculture, Fishery and Food (1971) reproduced below:

1.	Oesophagus rhabditiform	Free living nematode
	Oesophagus not rhabditiform	2
2.	Without sheath, oesophagus nearly half length of the body	*Strongyloides*
	With sheath, oesophagus less than 1/4th the length of the body	3
3.	Tail sheath short or medium length	4
	Tail of sheath very long	7
4.	Two refractile bodies or a bright transverse band visible between buccal cavity and oesophagus	*Cooperia*
	Refractile bodies or band absent	5
5.	Slender larva, tail of sheath of medium length tapering to a point and often kinked	*Haemonchus*
	Tail of sheath very short, conical	6
6.	Larva of medium size or large with distinct round tail	*Ostertagia*
	Small larva bearing one or two tuberosites or indistinctly rounded	*Trichostrongylus*

7	Very large larva, 8 gut cells, tail notched, bilobed or trilobed	*Nematodirus*
	Larva of medium size, 32 pentagonal gut cells, lumen of gut wavy	*Oesophagostomum*
	Larva of medium size, 32 square gut cells, lumen of gut straight	*Chabertia*
	Very small larva with 16 gut cells	*Bunostomum*

Pasture larval count

This technique is increasingly used in diagnosis of parasitic diseases in farm animals. A double 'W' route is followed on which 300–400 samples of grass are taken from the farm. Grass may be cut at soil level with scissors. In the laboratory, the grass is thoroughly soaked, washed and dried. The washings containing the larvae are passed through a sieve (aperture 38μm) tea strainer to remove the fine debris. Thereafter, centrifuge the washings to concentrate the larvae. The infective larvae are identified and counted microscopically under high power objective. The numbers present are expressed as L_3 per kg of dried herbage. The counts of more than 1000 L_3 per kg are considered moderately infected pastures where as count above 5000 L_3 per kg are expected to produce clinical disease in ruminants.

The infective larvae (L_3) of gastro-intestinal nematodes collected from faeces or pasture can be identified with the help of the standard keys.

Calculation

$$\text{Pasture larvae/ kg dry matter of pasture} = \frac{A \times C}{B \times D} \times 1000$$

Where

A = Volume of sediment (60 ml)

B = Volume examined (Mc Master) (0.6 ml)

C = Number of larvae obtained (total no. of larvae in 4 chambers)

D = Dry matter content of grass in grams

Micrometry

It is often important to know the size of the parasitic stages for identification. Micrometry is the measurement of microscopic organisms. In the late 16th century, Dutch scientist Antony van Leeuwenhoek, who used fine grains of sand as a gauge to determine the size of human erythrocytes, first reported the measurements with an optical microscope. Since then, numerous approaches have been employed for doing micrometry or morphometrics with the help of microscope. For performing the micrometry an ocular micrometer serves the purpose of scale or ruler. Ocular micrometer is simply a disc of glass which has etched lines on its surface. There are usually 100 equally etched spaced divisions, marked 0 to 10 upon an ocular micrometer. When placed in the eye piece, the ruled lines superimpose certain distance markers on the microscope field.

The actual value of one division of ocular micrometer is found by using a special microscopic glass slide known as stage micrometer. Stage micrometer has in its centre a known (one millimeter) distance etched into 100 equally spaced divisions. This 1 mm (1000 µm) distance is encircled and mounted by a cover glass. Thus each division of stage micrometer equals 0.01 mm or 10 micron. The distance of each division of stage micrometer becomes correspondingly enlarged under high power (10X and 40X) and oil immersion objectives (100X) of the microscope. Ocular micrometer is, therefore, calibrated under different powers of objective lens systems of the microscope. By determining how many divisions of ocular micrometer superimpose a known distance on the stage micrometer we can find out the exact value of one division of ocular micrometer in the microscope field. Once calibrated, the ocular micrometer can be used to measure the size of various parasites and eggs or oocysts.

In order to calibrate the ocular micrometry, the stage micrometer is placed upon the stage and the ocular micrometer in the tube, and arrange the two sets of rulings so that the first line in the ocular micrometer will coincide with the first line of the stage micrometer, and then find the value of one space in the ocular micrometer (Fig.7). To measure the diameter of a egg put the preparation on the stage, using the same objective and ocular micrometer, and note how many spaces a egg covers and multiply it with the value equal to 1 space of ocular micrometer.

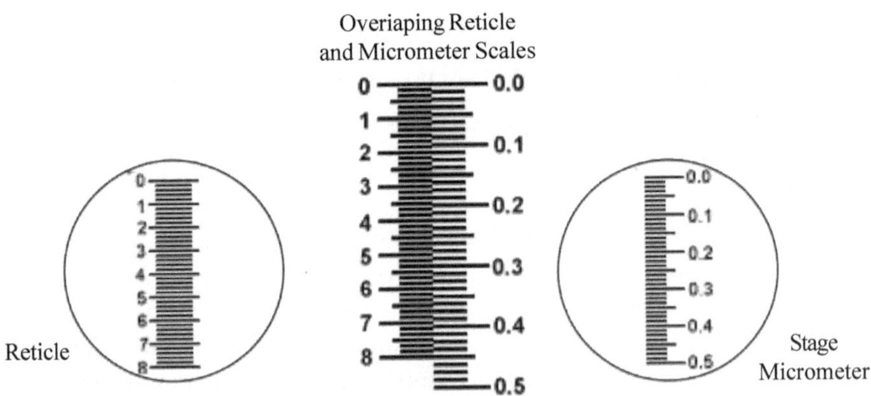

Fig. 7: Callibration of micrometer

3

Identification of Parasitic Mites Collection and Examination of Skin Scrapings

Mange or parasitic dermatitis is caused by the infestation of mange mites e.g. *Sarcoptes*, *Psoroptes*, *Demodex* spp. and a range of other mites. The condition is characterized by itching and rubbing of the infested parts of the body against hard objects. Lesions include hair loss, dry, rough, thickened and wrinkled skin accompanied by the presence of scab and foul odour in some cases. Some mites burrow into the host epidermis (*Sarcoptes*) while others spend their entire life cycle on the skin surface (*Psoroptes*). Most mites are microscopic or less than 1mm in length; therefore skin scrapings are taken from the affected area of animals suspected to be suffering from mange or parasitic dermatitis.

Collection of skin scrapings

Clip hairs in and around the affected skin areas of the animal showing mange lesions. The scrapings should preferably be taken from the periphery of the lesion with a scalpel, as the mites are more likely to be present on the periphery as compared to the centre of the lesions. Firstly, clean the scalpel blade by wiping it out with tissue paper and dip its blade in mineral oil and then scrape the patient's skin several times in one direction at the edge of the suspected area by pressing the lesion gently in between thumb and the index finger. Stop scraping when a slight amount of blood appears and transfer the scraping from the scalpel blade into a Petri dish, the edges of which are smeared with vaseline to prevent the escape of mites. For collecting samples suspected for demodicosis the affected area is squeezed between the fingers and the material oozing out should also be collected alongwith the scraping.

For detection of ear mites in dog, cat and rabbit, the animal should be gently restrained. A cotton swab is introduced into the external auditory canal and gently rotated. The swab is then placed on a piece of black paper and examined by means of a hand magnifier lens for detection of mites. An electrically illuminated otoscope cone may also be introduced directly into the ear canal for the detection of ear mites.

Precautions

- Before scraping, the lesions may be moistened with mineral oil or 10% KOH
- Scrape the skin deeply avoiding cutting of skin but some blood must ooze out
- Scraping so collected must be put in 10% KOH or NaOH in glycerine
- Dress the lesion properly after collection of scraping
- Avoid injury to auditory canal while collecting for ear mites

Examination of samples

The collected skin scraping can be examined directly either in water or saline or in light oil (clove oil) by placing the Petri dish containing it under the stereoscopic microscope. The movements of the mites if present can be enhanced by slightly warming the Petri dish on the flame. But if the collected material is too dense for direct examination, it should be brought just to boiling in 10% KOH or NaOH or the sample is kept for 24 to 48 hours to digest the skin debris so that examination of mite becomes easier.

The following methods are used for examination of digested sample:

Direct smear method

Transfer a few drops of the treated material on a glass slide, cover it with cover-slip and examine under low power of the microscope.

Sedimentation method

Transfer the treated material in a centrifuge tube and centrifuge it till sediment settles at the bottom. Discard the supernatant and make a smear from the sediment and cover it with a cover-slip. Examine it under low power of the microscope.

Sugar floatation method

Centrifuge the treated material in a centrifuge tube and discard the supernatant. Fill the centrifuge tube with sediment to half with distilled water and then upto top by Sheather's sugar solution and centrifuge it again. Mites come to the top due to higher specific gravity of the Sheather's sugar solution. Place a drop from the top to a glass slide, cover with cover-slip and examine under microscope.

Dispatch of samples

If mites or the skin scrapings are to be shipped to an acarologist or veterinary laboratory for identification, it should be stored in 70% ethyl or isopropyl alcohol. It should be sent in a tightly stoppered vial sealed with lac alongwith the informations regarding date and time of collection, host species, site of lesion on the host and the area or locality. The samples should preferably be sent through a messenger or by any quick mode of transport.

Techniques for diagnosis of demodectic mange mites

These mites are found either in the hair follicles as in dog or in skin nodules of certain animals (cattle, buffalo and pig). Hence, slightly different procedures are required to detect them:

For follicular mange

- Place a drop of mineral oil on the lesion
- Gently squeeze the skin while pressing it against a microscopic slide
- Place a cover-slip over the collected material and examine under a microscope

For pustular mange

- Hold the pustules between the thumb and the index finger, and press or incise the pustule, if hard
- A cheesy mass like a tooth-paste will come out which is collected in glass
- Transfer a small amount of material on a microscopic slide. Mix it with a drop of water and put a cover-slip
- Examine under a microscope

Identification of mange mites

Genus: *Demodex*

Common name: Follicular mite

Host: Man, Dog, Sheep, Goat, Cattle and Pig.

Demodex canis

- Occurs in the hair follicles and sebaceous glands of dogs
- These are elongate, cigar shaped and about 0.25 mm long
- Body divisible into head, thorax and abdomen, thorax bears four pairs of stumpy legs, abdomen is elongated and transversely striated on the dorsal and ventral surface

- Mouth parts consist of paired palps and chelicerae and a single hypostome
- Eggs spindle shaped and are laid in hair follicle
- Larvae six-legged, nymph and adult with four pair of legs

Genus: *Sarcoptes*

Common name: Scabies mite

Host : Man, Sheep, Goat, Cattle, Pig and Equine

Sarcoptes scabiei

- Small in size, female 0.33-0.6mm x 0.25-0.4mm, male 0.2-0.24 × 0.15-0.2mm
- Body roughly circular, dorsal surface covered with fine fold and grooves and bears numerous small triangular scales
- Legs are short and grouped in two pairs on either side of nymph and adult and they do not project beyond the margin of the body
- The tarsi of the Ist, IInd and IVth pair of legs in males and Ist and IInd pair of legs in females end in bell shaped suckers (caruncles), others ending in bristles
- Eggs are oval and laid in tunnels burrowed by females
- Larvae are six-legged, nymph and adult with four pair of legs

Genus: *Notoedres*

- Body globose, covered with scales or spines
- Legs short, pedicels unsegmented as in *Sarcoptes*
- Anus dorsal, while in *Sarcoptes* it is terminal

Notoedres cati: Cat

Notoedres muris: Rodents

Genus: *Cnemidocoptes*

- Body globose, dorsal surface devoid of spines
- Legs short and stumpy, all legs of males possess suckers while none of the legs of female have suckers

Cnemidocoptes gallinae

Common name : Depluming itch mite

Host : Fowl

- The mite burrows into the skin along shafts of the feathers of back and wings

Cnemidocoptes mutans

Common name: Scaly leg mite of poultry

Host: Fowl and Turkey

- The mites get on to the feet of the birds from the ground and lesions develop from toes upwards, may lead to lameness and malformation of feet

Genus: *Psoroptes*

Common name: Sheep scab mite

Host: Sheep and Cattle

Psoroptes ovis

- Oval in shape and larger than *Sarcoptes*
- All legs in nymphs and adults are long and extend beyond the margin of body
- The tarsi of the Ist, IInd and IIIrd pair of legs in males and Ist, IInd and IVth pair of legs in females end in bell shaped suckers
- Dorsal surface is devoid of scales and spines but the cuticle shows very fine striations

P. equi: Horse, Donkey, Mule

P. natalensis: Cattle, Zebu, Buffaloes

P. cuniculi: Ears of Rabbit, Goat, Sheep, Horse

Genus: *Chorioptes*

Common name: Foot mange or itchy leg mite

Host: Cattle, Horse, Sheep and Goat

Chorioptes bovis

- Resembles *Psoroptes* except tarsi of all legs of males and Ist, IInd and IVth pair of legs in females end in bell shaped suckers

Genus: *Otodectes*

Common name: Ear mange mite of dogs

Host: Dog, Cat and other wild carnivores

Otodectes cynotis

- It resembles the *Chorioptes* mite

Genus: *Cytodites*

Cytodites nudus

Common name: Air sac mite of poultry

- Small, oval creamy coloured mite with suckered legs
- Lives in the respiratory passages and air sacs

Identification of ticks

Family: Argasidae (Soft ticks)

Genus: *Argas*

Argas persicus

Common name : Fowl tick or blue bug

Host : Fowl, Turkey, Pigeon, Duck and wild birds

- Oval in shape, narrow anteriorly than posteriorly
- Engorged tick has slaty blue colour
- Mouth parts sub-terminal and not visible from dorsal aspect
- Sexual dimorphism is not marked
- Eyes are absent
- Dorsal shield or scutum absent

Economic importance: Transmits *Anaplasma marginale, Aegyptianella pullorum, Borrelia anserina* (fowl spirochaetosis)

Otobius megnini

Common name: Spinose ear tick

Host : Dog, Sheep, Goat, Cattle, Horse, Pig and Cat.

- Nymph and larvae are parasitic
- The nymphs are widest at the middle and skin is mammillated and bears numerous spines like projections
- Body colour of nymph is bluish grey, while legs, mouthparts and spines are pale yellow

- Adults that are not parasitic have a constriction at the middle, giving the body a fiddle shape

Hard ticks

- Ticks possess hard chitinous shield or scutum, which extends over the whole dorsal surface of the male and covers only a small portion behind the head in larvae, nymph and adult female
- Mouth parts anterior and visible from dorsal aspect
- Adults have one pair of spiracles situated postero-laterally on the 4th coxae
- Posterior border of the body may be notched forming festoons
- Some ticks may have coloured, enamel like areas on the body i.e. ornate ticks

Genus: *Ixodes*

- Inornate ticks
- Eyes and festoons absent
- Palps and hypostome long (longirostrate)
- Anal groove surrounds the anus anteriorly (prostriata)

Genus: *Rhipicephalus (Boophilus) microplus*

- Inornate ticks
- Eyes present, festoons absent
- Palps and hypostome short (brevirostrate)
- Anal groove absent in females and faint in male and surrounds the anus posteriorly
- Coxa one bifid, caudal process prominent in males
- Spiracles circular or oval

Boophilus microplus: Transmits *Babesia bigemina* (red water disease) and *Anaplasma marginale*

Genus: *Hyalomma*

- Inornate ticks sometime ornate
- Eyes present, festoons present or absent
- Palps and hypostome long (longirostrate)

- Males with a pair of adanal shields
- Spiracles comma shaped in male and triangular in females

Hyalomma anatolicum anatolicum: Transmits *Theileria annulata* (bovine tropical theileriosis), *Anaplasma marginale*.

Genus: *Rhipicephalus*

- Usually inornate ticks
- Eyes and festoons present
- Palps and hypostome short (brevirostrate)
- Basis capituli hexagonal dorsally
- Coxa I with two strong spur
- Spiracles comma shaped, short in females and long in males

Rhipicephalus sanguineus: Commonly called brown dog tick, transmits *Babesia canis, Babesia gibsoni, Ehrlichia canis* and *Hepatozoon canis.*

Genus: *Haemaphysalis*

- Inornate tick
- Eyes absent, festoons present
- Palps short and conical
- Trochanter of first pair of leg bears a dorsal process
- Spiracles in female ovoid or comma shaped, in males are ovoid

Genus: *Dermacentor*

- Usually ornate
- Eyes and festoons present
- Palps and hypostome short
- Coxa I bifid, coxa IV of male much larger than coxae I to III

Genus: *Amblyomma*

- Usually ornate
- Eyes and festoons present
- Palps and hypostome long
- Male without ventral plate

4

Examination of Blood, Lymph Node Biopsy and Other Body Fluids

Collection of blood

Blood sample should be collected in a capped clean and dry vial containing an appropriate amount of anticoagulant. The commonly used anticoagulants are ethylene diamine tetra acetic acid (EDTA) @ 1-2 mg/ml and heparin @ 5 I.U./ ml of blood.

Sites for blood collection

Before collection of blood clean the site of collection by applying some antiseptic solution. If the area is hairy shaving of the site under aseptic condition should be done. **Tip of the ear should be preferred for prompt or early diagnosis of haemoprotozoan infections**. The vein can then be punctured with hypodermic needle and the collected blood is transferred to a clean glass vial. For collection of blood various preferred sites in different animals are as under:

Jugular vein	:	Cattle, buffalo, sheep, goat and horse
Cephalic vein	:	Dog
Wing vein	:	Poultry
Ear vein	:	Cat, cattle, buffalo, sheep, goat and horse
Anterior vena cava	:	Pig

The collected blood should be sent to laboratory at the earliest under cool conditions in a thermos flask containing ice. The blood vial must be tightly capped and covered with adhesive tape to prevent entry of water into the vial during transportation. Methanol fixed blood smear can also be sent for examination after wrapping it in a piece of paper or in a slide mailer. Before dispatching, identification mark on the slide and sample should be provided. Detailed information regarding the history of the case, address of owner, preservative used and treatment given must be sent with the sample. At the end it must be signed by Veterinary Officer/dispatcher along with designation, address and seal.

Requirements for the blood smears

Slides: Clean the new glass slides by dipping overnight in absolute alcohol and then dry with tissue paper or clean muslin cloth to remove the grease coating from the slides. Alternatively, glacial acetic acid (2%) can also be used for cleaning of slides.

Stains

(a) Giemsa Stain (b) Leishman stain (c) Wright's stain

Staining jars and racks

Buffer tablets/capsules: pH=6.8-7.2

Slide warming table

Types of blood smears

Wet blood film

This is an easy method used for quick detection of haemoflagellates like trypanosomes and microfilariae of some nematodes e.g. *Dirofilaria immitis*, *Dipetalonema reconditum.*

Procedure

Put a drop of blood on the centre of microscopic slide, put a cover-slip over it and first examine at 10X and then at 40X under microscope.

Thin blood smear

- Place a small drop of blood on one end of the slide
- With the help of another slide, spread the blood keeping the angle in between 30° and 45° and push it quickly so that an uniform thin blood film is made (In the microscope the erythrocytes should be seen lying "shoulder to shoulder")
- Air dry the blood smear quickly
- Fix the smear in methanol for 1-2 min and allow the alcohol to evaporate and if you are using ethanol, fix it for 15 minutes
- Thin blood smears help in the study of morphology of parasites

Thick blood smear

Such smears enable us to examine more blood in a limited area and in a short period. Thick smear is made for detection of chronic infections, as the number of parasites in peripheral circulation is low and thin smears are unable to reveal the infection. These types of smears are not used for avian or camel blood because of their nucleated red blood cells. These smears are helpful in detection of intercellular parasites, after dehaemoglobinization and staining.

Procedure

- Place a large drop of blood in center of a clean slide
- Spread it with the corner of other slide in an area of about 1 cm diameter
- Air dry the thick smear thoroughly by covering under a Petri dish so as to protect it from dust and flies
- The smear is dehaemoglobinized by placing it in distilled water until the colour has disappeared
- Fix the smear and stain it

Staining procedures

Giemsa stain

- A 10% Giemsa stain is prepared by diluting Giemsa stain stock solution (1:9) in phosphate buffer (pH 6.8-7.2)
- For the best staining result, the prefixed slides are stained for 45 minutes in a jar in which the slides are dipped completely or by keeping the slides on staining rack and then layering them with diluted stain solution
- The slides are removed from the staining jar/staining rack and washed twice with distilled water or buffered solution
- After washing the slides are air dried
- The stained smears are examined under a microscope using 40X magnification (for microfilaria) and 100X/oil immersion (for haemoprotozoan)

Wright's stain

- Prepare thin blood smear and air dry it
- Put few drops of undiluted stain on smear and allow it to act for 1 to 2 minute
- Add equal number of drops of phosphate buffer and mix gently without touching the smear

- Allow the diluted stain to act for 2 to 4 minutes
- Wash the stained smear with neutral buffer or tap water
- Air dry it and examine under 40X or 100X magnification of microscope.

Leishman stain

- Prepare thin blood smear and air dry it
- Add few drops of undiluted stain and allow it to remain for 1 to 2 minutes
- Dilute the stain on the smear with the double amount of phosphate buffer and mix gently
- Allow the diluted stain to act for 10 to 15 minutes
- Wash the stained smear with distilled water to drain of the excess stain
- Examine the smear under 40X or 100X magnification of microscope

The well stained blood smears will reveal the colour of various cells as under:

Nucleus of lymphocyte	:	Deep purple/violet
Cytoplasm of lymphocyte	:	Clear blue
Nucleus of monocyte	:	Red purple
Cytoplasm of monocyte	:	Blue
Granules of eosinophils	:	Brick red
Granules of neutrophils	:	Colourless
Granules of basophils	:	Blue
Erythrocyte	:	Light pink to bluish (based on pH)

Acridine Orange Stain

Procedure

1. Fix the blood smear in methanol for 1-2 min
2. Then pour the 0.01% Acridine orange stain solution on the blood smear
3. Allow it to act for 2-3 min
4. Rinse in water, slowly (preferably at pH around 3.5-4)
5. Put a coverslip immediatedly
6. Examine immediatedly in fluorescent microscope with blue filter

Result: Parasites takes yellow orange colour and host cell nucleous takes green fluorescent only

Advantage: Rapid, quick, sensitive for diagnosis of Rickettsial parasites, (*Anaplasma*, *Ehrlichia*) and also used for *Theileria*, *Babesia* & malaria parasites diagnosis.

Disadvantage: It need fluorescent microscope

Buffy coat examination

Buffy coat is examined for detection of microfilariae of heart worm of dog (*Dirofilaria immits*), *Trypanosoma evansi*, *Ehrlichia* spp., *Babesia* spp. and *Hepatozoon canis*.

Procedure

- Blood is taken in a centrifuge/microhematocrit tube
- It is then centrifuged at 3000 rpm for 5 minutes
- The centrifuged tube has three layers:

Upper (yellow layer) : Plasma

Middle (white layer) : WBCs

Bottom (red layer) : RBCs

- Discard the plasma layer with the help of pipette or syringe
- Take the buffy coat and upper most layer of RBCs/packed cells
- Put it on clean glass slide, draw thin smear and stain it with Giemsa stain
- Examine the smear under 40X or 100X magnification of microscope

Identification of haemoprotozoan parasites

Trypanosoma evansi

Host : **Domestic and wild animals**

Location : Blood, plasma and tissue fluids

Disease : '**Surra**' in animals

Trypomastigote

- Fusiform shaped body, medium size, monomorphic; 15-34 µm in length
- Nucleus centrally placed

- Undulating membrane well developed and free flagellum prominent
- Posterior extremity is short and frequently truncated
- Kinetoplast is sub terminal and placed at posterior end

Leishmania donovani

Host : **Man (and Dog)**

Vector : *Phlebotomus argentipes*

Disease : Visceral leishmaniosis, Kala-azar, Black fever, Tropical Splenomegaly, Dum Dum fever

Amastigote

- Seen in cells of reticulo endothelial system of the host (macrophages)
- Ovoid shape with posteriorly placed nucleus
- 8-10 amastigotes in a host cell cytoplasm
- Kinetoplast anterior to nucleus
- Axoneme short and rod shaped

Plasmodium gallinaceum

Host : **Domestic fowl and other birds**

Vector : *Culex* and *Aedes* sp. of mosquitoes

Location : Erythrocytes

Trophozoites

- Ring like appearance in the RBC (Signet ring)
- Large vacuole
- Chromatin at the peripheral

Schizonts (Size 8x5 µm)

- Irregular mass with varying sized in the cytoplasm having numerous nuclear fragments
- Presence of pigment granules

Gamonts

- The gamonts in red blood cells are seen round and possess few pigment granules of large size
- Gamonts displace host cell nucleus

Haemoproteus columbae

Host : **Pigeon**

Vector : Pigeon fly - *Pseudolynchia canariensis*

Gamonts

- Elongate and sausage shaped bodies which partially embrace the nucleus of the erythrocytes
- They contain a variable dark brown pigment
- Microgametocytes (male) stain pale; the nucleus is pale pink and diffuse and pigment granules are blue collected into a spherical mass
- Microgametocytes (female) stain dark blue; the nucleus is dark purple to red and compact
- Pigment granules are dispersed throughout the cytoplasm

Hepatozoon canis

Host : **Dog**

Vector : Brown dog tick - *Rhipicephalus sanguineus*

Schizonts

- Asexual stages occur in the endothelial cells of the spleen, bone marrow and liver
- Several generations of schizonts occur
- The latter enter polymorphonuclear leucocytes and develop to gamonts

Gamonts

Location : Neutrophils

- These are elongate or rectangular or gelatin capsule shaped bodies of

8-12 X 3-6 μm (average 6X3 μm) surrounded by the delicate capsule

- Nucleus of the leukocyte is displaced
- Nucleus of gamont is compact, with chromatin situated at one end as horse-shoe-shaped

Babesia bigemina

Host : **Cattle and Buffalo**

Disease : Cattle tick fever, Red water fever, Bovine Babesiosis, Piroplasmosis

Vector : *Rhipicephalus (B.) microplus* and *R. (B.) annulatus*

Morphology

- Large form of piroplasm (4.5 x 2 μm) in red blood cells
- A pair of pear shaped forms (merozoites) in red blood cells form an acute angle at the narrow end
- A string of chromatin material present along the narrow end
- Single or many round or amoeboid forms may also be found

Babesia canis

Host : **Dog**

Vector : Brown dog tick - *Rhipicephalus sanguineus*

Disease : Biliary fever, Malignant Jaundice and Canine Babesiosis

Morphology

- Large form of piroplasm
- Merozoites pyriform in shape, 4-5 μm in length
- Multiple infections are common in a single erythrocyte - 2, 4 or 16 organisms may be found in a single erythrocyte
- Pleomorphic, amoeboid to ring forms

Babesia gibsoni

Host : **Dog**

Vector : Brown dog tick - *Rhipicephalus sanguineus*

Morphology

- Small form of piroplasm
- Pleomorphic, usually merozoites lack the usual pyriform shape
- They are typically annular or oval, signet ring, rarely ovoid to circular forms

Babesia caballi

Host : **Horses**

Vector : *Dermacentor reticulatus, Rhipicephalus bursa*

Morphology

- Large form of piroplasm like *B. bigemina*
- Pyriform in shape (2.5 to 4 µm in length)
- Round or oval forms (1.5 - 3 µm in diameter) may also occur

Babesia equi

Host : **Horse**

Vector : *Hyalomma* sp., *Rhipicephalus* sp. and *Dermacentor* sp.

Morphology

- Small form of piroplasm 1-3 µm in length
- Pyriform trophozoite divides into four daughter organisms (in tetrads) which frequently form '**Maltese cross'**
- Pear shaped, round, amoeboid or annular forms may also occur

Theileria annulata

Host : **Cattle, Buffalos**

Disease : **Tropical Bovine Theileriosis**

Vector : *Hyalomma anatolicum* and *Hyalomma marginatum isaaci*

Morphology

- The erythrocytic forms are highly pleomorphic - ring, oval, round (0.5 - 1.5 µm), rod or comma shaped (1.6 µm)
- Small quantity of cytoplasm is seen in oval and ring forms as a thin ring

around the central vacuole with a cap like chromatin

- One or more organisms are found in one red blood cell

Schizonts (Koch's blue bodies - KBB)

- Found in lymphocytes (spleen and pre scapular lymph nodes)
- Appear as irregular or rounded bodies with blue matrix, containing numerous pinkish chromatin particles in the cytoplasm of the lymphocytes
- They may be found free in the blood smear due to rupture of host cell

Anaplasma marginale

Host : **Cattle, Buffalo, Sheep and Goats**

Location : Erythrocytes

Vector : Several species of ticks and biting flies

Disease : Anaplasmosis / Gall sickness

Morphology

- Oval or spherical bodies (0.2 - 0.5 μm in diameter)
- One or two organisms may be found mostly at the periphery of red blood cells
- Organisms appear as dense homogeneous red to dark red in colour with Romanowsky stains
- A faint halo may be seen around anaplasma-bodies

Ehrlichia canis

Host : **Dog**

Vector : *Rhipicephalus sanguineus*

Disease : Canine ehrlichiosis, Tropical pancytopaenia and Canine typhus

Morphology

- Found within the cytoplasm of mononuclear cells
- Elementary bodies are small, gram-negative organisms about 0.5 μm in diameter
- Infected monocytes contain several morulae and each morula containing several elementary bodies

Xenodiagnosis

In which feeding of triatomids bugs on the suspect or allowing bugs to feed on the patient's blood through a membrane, may be valuable techniques. Laboratory reared and *T. cruzi* free triatomes are used and after 2-4 weeks and if the suspected material is +ve, metacyclic trypanosomes are found 7-10 days later in the dropping of the bug.

Examination of biopsy material for protozoa

- Biopsy can be very helpful in the diagnosis of various protozoan infections.

Visceral leishmaniosis

- The only certain method of diagnosis is the demonstration of the organism in spleen pulp, lymph nodes and bone marrow
- Fine needle aspiration of the spleen for culture and touch preparations is the most sensitive methods, yielding a diagnosis of visceral leishmaniosis in 96-98% of cases
- The procedure is reasonably safe provided that it is done by experienced operator and the patient has normal coagulation profile
- Examination of bone marrow aspirates results in a diagnosis in more than half of the cases. An alternative approach is the biopsy of the liver or aspiration of the enlarged lymph nodes
- In patients with concurrent HIV infection, amastigote may be found in unexpected sites such as bronchoalveolar lavage fluid or pleural effusion or in biopsy specimen of mucosal lesions in the oropharynx, larynx, stomach or intestine
- Sternal puncture is increasingly used in diagnosis
- In the Dog, scraping should be made from the periphery of the skin ulcers or eczematous areas

Theileriosis

- Under general field conditions, the most satisfactory diagnosis is made by the demonstration of the schizonts in material obtained from the superficial lymph nodes or by spleen puncture
- But, differentiation between *T. parva* and *T. annulata* is difficult on the basis of the biopsy examination and the differential diagnosis is based on the enzootic area, pathology, epidemiology and preferably, cross immunity tests

***Sarcocystis* spp.**

Host : **Cattle, Buffalo, Pigs and other animals**

Sarcocysts / Miescher's tubules

- Sarcocysts are either macroscopic or microscopic
- These may be filamentous, globular, elongate, ovoid or cylindrical in shape located in between the muscle fibers
- There are a number of septa in between the cysts containing numerous bradyzoites
- The structure of the cyst wall is a useful criterion for speciation

Bradyzoites

- Present in the sarcocysts
- Crescent or sickle shaped bodies with one end blunt and the other end pointed
- Nucleus is placed near the blunt end

Toxoplasma gondii

Host : Domestic Cats, Rats, Mice and many other warm-blooded animals

Intermediate Host: Man, domestic animals and birds

Disease : **Storming abortion** in sheep

Tachyzoites (Fast multiplying forms)

- Found freely in blood and other tissue and body fluids in 'clones' or groups or 'pseudocysts'
- Cresent shaped bodies with one end blunt and the other pointed
- Nucleus is centrally placed

Bradyzoites Tissue cysts - (slow multiplying form)

- Microscopic cysts with thin wall (older ones measure 100 um in diameter)
- Spherical within the host cell of visceral organs (lungs, liver and kidney), neural and muscular tissues, including the brain, eye, skeletal and cardiac muscles
- Numerous bradyzoites are found within each tissue cysts

Neospora caninum

Host : Dog

Intermediate host : Cattle, Sheep, Goats and Horses and parasites can be transmitted to mice, cattle, rats, gerbils and sheep experimentally

Disease : Ascending paralysis in young pups, Abortion in infected cow

Tachyzoites

- Found in host cell cytoplasm i.e. neural cells, macrophages, fibroblasts, vascular endothelial cells, myocytes, renal tubular epithelial cells, hepatocytes
- Ovoid, lumate, globular 3-7μm x1-5μm

Bradyzoites Tissue cyst

- Round to oval and up to 107μm long
- Found only in central nervous systems with one exception in retinas of congenitally infected dogs
- Cyst wall is upto 4 μm thickand enclosed upto 200 bradyzoites of 6-8 μm x1-1.8μm in size

Lymph node impression smears

Lymph node smears are prepared for the confirmation of theileriosis (*Theileria annulata*) and equine babesiosis (*Theileria equi*). Smears prepared from positive cases would reveal the presence of asexual stage (schizonts). The lymph nodes generally preferred are prescapular and precrural. A whole lymph node or a part of it is collected from dead animal.

Procedure

In case of dead animal

- Cut a portion of the lymphatic tissue with a sharp knife
- The cut surface is touched on a clean glass slide at two or three places about half a centimeter apart and many preparations are made
- Smears thus made are air dried, fixed in methanol and stained with Giemsa solution

In case the live animal

- Smears are made from lymph gland aspirate

- Site of the suitable lymph gland is shaved and cleaned with a swab of alcohol/spirit
- The gland is punctured and small amount of sterile normal saline solution (0.5-1.0 ml) is injected with a sterilized hypodermic needle attached to a syringe.
- The plunger of the syringe is withdrawn after some time to suck the lymph node aspirate and little amount is ejected onto a clean glass slide.
- The smears are prepared just as blood smears and stained after fixation with Giemsa staining solution.

Interpretation

In the positive cases of *Theileria* infection in cattle lymphocytes enlarged showing macroschizont (Koch blue bodies) which is intracellular, circular or irregular pleomorphic, blue cytoplasm having varying number of red chromatin granules or developing merozoites from the lymph node and spleen etc.

The two types of schizonts are found under Giemsa stain viz., macroschizonts is 8 μm diameter which contain up to eight developing merozoites and microschizont, similar size which contains up to 36 developing merozoites.

Mice inoculation test

The mice inoculation test is useful in detecting subclinical infections in cattle, buffalo and camel or where the blood smears do not show blood protozoa especially trypanosomes in cattle and buffaloes as it is more sensitive than the direct microscopic examination.

Procedure

- Fresh blood from suspected animal is inoculated into albino mice (Fig 8), rats, guinea pigs or rabbits. Albino mice are probably the best experimental animals as they are highly susceptible and easy to handle
- About 0.5 ml of blood is injected intra-peritoneally in mice and is examined after 24 h interval for presence of parasites
- In case of trypanosomosis the protozoa appear in circulation, two to three days after inoculation in mice while in other animals like rats, rabbits and guinea pig parasites appear in peripheral blood in one to two weeks.

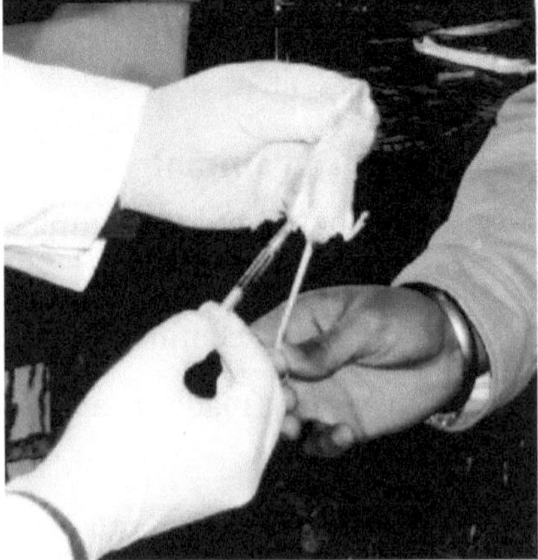

Fig. 8: Mice inoculation test

- Tail blood is collected from mice after snipping the tail end with sharp scissor and is put on microscopic slide for detection of trypanosomes

Examination for *Tritrichomonas foetus*

Examination of preputial washings for *Tritrichomonas foetus* in male

- Restrain the animal properly and inject about 50 -100 ml of sterile saline solution into the preputial cavity with the help of catheter
- Massage the prepuce and cauterize the mucus of prepuce
- Draw back the injected saline with the help of syringe and collect it in a vial
- Centrifuge the irrigated material at 1000-1500 rpm for five minutes and examine the sediment, or allow the material to stand for 1 to 2 hours to settle the organism. Discard the supernatant and take a drop of sediment on the slide for presence of the trichomonad which gives typical jerky movement under low power of microscope
- In very light infection the smear from the sediment is stained with the Giemsa stain to yield good results. The bull is considered free of infection when six consecutive tests at weekly interval are negative and two virgin heifers bred free from it show no infection

Examination of vaginal or uterine discharge for *Tritrichomonas foetus* in female

- In females, vaginal mucus or exudates, uterine discharge especially two to three days before the expected time of next oestrus or after abortion can be examined for best results
- Samples can be collected by washing the vagina with saline or by irrigation of the uterus
- In females, at least three negative test and two normal oestrus periods indicate cow free from trichomonad infection
- In aborted fetuses, trichomonads are found in the mouth cavity and stomach contents along with amniotic and allantoic fluids. Organisms disappear after 48 hours of abortion, so care should be taken that no contamination occurs with faeces
- In heavy infections the organism can be demonstrated in purulent mucus by direct examination on the glass slide under microscope while in lighter infections the sediment should be examined after centrifugation. However, the sample should be kept under warm conditions throughout and just

before examination warm saline can be added to observe the characteristic motility of the parasite

- Preputial contents, vaginal and uterine discharges, aborted foetal stomach content if positive can be cultured in Modified Diamond's trichomonad medium, Oxoid's *Trichomonas* medium, InPouch TF test (Biomed Diagnostics) within 24 hours of collection. Media are incubated at 37°C for 7 days and is examined daily for the presence of *T. foetus*.

Identification of Tritrichomonas foetus

Host : **Cattle**

Location : Preputial cavity in bull and vaginal cavity in cows

Disease : Bovine trichomonosis, Early abortion, permanent sterility in cows

Morphology

- Pear shaped and size is 10-25μm x 3-15μm
- Three anterior flagella and one posterior free flagellum
- Prominent axostyle, well developed undulating membrane
- Nucleus placed anteriorly
- Cost is prominent
- Parabasal body is sausage shaped

***In vitro* cultivation of the protozoa**

In vitro cultivation of the parasitic protozoa may be helpful in research on basic biology of the protozoa. Recently, commercial availability of the standard culture media, sterile plastic wares and wide spectrum antibiotics have greatly simplified the establishment of basic culture techniques. *In vitro* culture of protozoa is advantageous in terms of use of laboratory animals can be reduced for routine laboratory maintenance of protozoa, host-parasite interaction studies on physiological and biochemical aspects and preliminary screening of anti-protozoal drugs.

The primary aim of the culture is to provide artificial environment in which parasites show sufficient growth outside the host, under controlled condition. Trypanosomes were among the first parasitic protozoa to be cultivated *in vitro*. Axenic culture is essential for those organisms as these will not grow successfully *in vitro* in the presence of bacterial and fungal contaminants. To discourage the growth of bacteria, antibiotics (penicillin and streptomycin) are frequently used which do not have significant harmful effects on most flagellates. The cultivation of the pathogenic protozoa parasitic in the blood of the vertebrates has not shown sufficient high level of development e.g. *Plasmodium* has been maintained *in vitro* only within their living host cells. Major advances in the field of cultivation of blood and tissue stages of parasitic protozoa have occurred in recent years. Now all or parts of their life cycle of *Trypanosoma, Leishmania* and *Plasmodium* can be cultured *in vitro*. A variety of media have been developed in recent years viz.,

Media	Parasite grow
Dobell's 'Hsre' medium: Most reliable media used for the growth of parasites in axenic and monoxenic culture	Consists of solid phase in the form of heat-coagulated protein (horse-serum) and liquid phase as egg-albumin diluted with normal saline solution. For Carbohydrate source, a loop full of sterile rice is added in the medium.
Boeck-Drbohlav diphasic system	*Entamoebae, Trichomonas* and *B.coli.*
Axenic and monoxenic complex media 1. TYI S 33 (Trypticase-yeast –iron-serum) media 2. TYM (Trypticase-yeast-maltose) medium	*E.histolytica, E.coli, E.invadens, E.terrapine, E.moshkovskii, Dientamoebae fragilis, E.hartmanii, B.coli* and *Pentatrichomonas hominis*. (For the growth of *E.gingivalis* dextran I is added instead of glucose). *Trichomonas hominis, T.vaginalis* and *T.foetus* (For culture of *T.tenax* and *T.muris* organisms, addition of a mouse caecal extract is needed).
Balamuth's amoeba medium	*Amoebae*
Beef extract glucose peptone serum (BGPS)	*Trichomonas*
Ringers's serum starch medium (RSSM).	*B.coli*
NNN (Novy, McNeal and Nicolle) medium	*Leishmania*

Examination of blood for detection of microfilariae

Microfilaria is the larval stage of the filarial nematode present in the blood or lymph and can be diagnosed by employing the following methods:

Wet blood film

In heavy infection, take a drop of fresh blood and place it in the center of a clean microscopic slide. Cover it with a cover-slip and examine immediately under the low power (10X) of the microscope. Look for the slow serpentine like movements of nematode larvae. The larvae may retain their motility for as long as 24 hours on the microscope slide.

Filter method

- Take 1 ml or more of blood in sodium citrate or in any other anticoagulant and lyse it with 0.1% sodium carbonate

- Lysed blood is then filtered through a filter paper of 3 µm porosity
- Microfilariae are retained on the filter paper and are stained with 0.1% methylene blue and then examined under microscope (40X)

Knott's method

- Take 1 ml of heparinised blood in a centrifuge tube and mix it with 9 ml of 2% formalin
- Centrifuge for five minutes at 500 rpm
- Discard the supernatant
- Stain the sediment with equal volume of 1:1000 dilution of methylene blue
- Place a drop of mixture with the help of pipette on a glass slide, put a cover-slip on it and examine under microscope

Interpretation

Microfilariae are stained blue and are easily identified.

Characteristic for differentiation of microfilariae

Microfilaria of the pathogenic heartworm (*Dirofilaria immitis*) and of other, less pathogenic filarioid species (*Dipetalonema reconditum*) can be distinguished on the basis of several criteria like size, shape, movement, numbers and staining characteristics:

Characteristic	*Dirofilaria immitis*	*Dipetalonema reconditum*
Body shape	Usually straight	Usually curved
Body length	295-325 µm (>300 µm)	250-288 µm (<300 µm)
Mid body width	5-7.5 µm	4.5-5.5 µm
Shape of cranial end	Tapered	Blunt
Shape of tail	Straight	Usually curved or hooked
Motility pattern	Tend to remain in the general area and coil and uncoil.	Move rapidly across the microscopic field with serpentine motility

Examination of urine

The urine sample may be collected during normal urination and the owners may be instructed to collect the samples at home in a clean container. The sample should be properly labelled and kept in a refrigerator until examination. The sample is centrifuged and the sediment is utilized for the microscopic examination at low power after putting on the clean microscopic slide and placing a cover-slip over it. Common parasitic stages encountered in urine are:

- Eggs of *Dioctophyma renale* (Dog)
- Eggs of *Stephanurus dentatus* (Pig)
- Eggs of *Capillaria* spp. (Dog)

Examination of nasal scrapings

It is conducted for the diagnosis of *Schistosoma nasalis* infection in cattle and buffaloes. A clean, sterile spatula is inserted in the nasal cavity after restraining of the animal and is subsequently rotated for collection of the nasal scrapings in 10% formalin. The material is examined after centrifugation and typical boomerang shaped eggs are seen in positive cases.

Examination of sputum/nasal discharge

It is done for the detection of eggs and larvae of parasites present in lungs. A drop of the sputum or nasal discharge is put on a glass slide and examined under the microscope. Eggs of *Paragonimus westermanii*, larvae of ascarid worms and eggs containing larvae of lungworms may be seen in positive cases.

5

Collection and Preservation of Parasites

HELMINTHS

Collection of parasites

Fresh helminth parasites should be collected, as far as possible to achieve better results in staining. This can best be done by conducting a post-mortem examination of the animal for which the procedure given below should be followed. On opening the body and before removal of the viscera, examine the body cavity (for free lying parasites, e.g. *Setaria*), subcutaneous tissue (for *Parafilaria* and *Hypoderma* larvae) and the surfaces of internal organs for spot or nodules that might contain parasites. Remove them for histopathological examination (in 10 % formalin). Remove the visceral organs separately and place them in different enamel trays.

Digestive tract: Cut various portions such as oesophagus, stomach (or rumen, reticulum, abomasum), small intestine, colon, caecum and rectum. Each portion should be tied with thread on both ends before cutting to prevent its contents mixing with those of other parts. Incise each portion length wise and dislodge its contents by flushing it with water into the tray. Hold the cleaned portion against light (electric lamp) for parasites embedded in the mucosa (*Strongyloides papillosus* of cattle) or in nodules (*Spirocerca lupi* of dog). If present, remove those parts and put them in 10% formalin for future investigations or histopathological examination. Each part should now be scraped with a scalpel down to the muscular layer and scrapings are collected in the same tray with intestinal contents. Stir the suspension of intestinal contents and scrapings thoroughly and allow it to stand for some time (15 minutes). This process can be repeated by adding water and decanting the supernatant but without disturbing the sediment. The parasites visible to the naked eye should be picked up from the sediment with a hair-brush or a crooked needle.

Liver: It should be incised first along the bile duct as this will allow the parasites present in bile ducts (*Fasciola gigantica*) to escape out into the tray. Next it should be torn apart or sliced in water and the suspension examined under a dissecting microscope or with a magnifying glass for smaller parasites (*Dicrocoelium* and *Opisthorchis*).

Gall-Bladder: Along with the liver, this organ should also be examined by opening it and its contents (bile). In addition, inside of the bladder may also be examined with naked eye or with a magnifying glass for parasites (*Gigantocotyle explanatum*).

Kidney: This organ should be examined much in the same way as the liver by slicing it in normal saline, thus allowing the parasites or their larval stages to escape out into the tray. When a parasite is large (*Stephanurus dentatus* of pig), it can be seen with naked eye.

Lungs: Open and examine lungs alongwith trachea, bronchi and alveoli for lung worms or larval stages of other parasites.

Heart and blood vessels: Cut open and search their insides with naked eye for parasites (*Dirofilaria immitis* or *Onchocerca armillata*). Helminths (schistosomes) present in mesenteric/portal veins may be detected by stretching the mesentery against a light source. Parasites, if present, can be collected by incising the vein and allowing the blood alongwith the flukes to flow into a Petri dish or a tray.

Other organs: Organs/tissues/cavities like eyes (*Thelazia*), nasal passages (*Schistosoma nasalis*), and reproductive organs like ovary and oviduct (*Prosthogonimus*) should also be examined.

Skeletal muscles: Skeletal muscles (diaphragm, etc.) are treated for harvesting parasites from them. Two methods are generally used:

A. **Compression method**: It can be used for revealing the larvae of *Trichinella spiralis* or cyst of *Sarcocystis*. Place a thin slice of muscle between two slides. Press hard to flatten the muscle and examine under low power of a microscope.

B. **Digestion method**: Effective for collection of larvae in the tissues. Place minced muscle (or other tissue like liver) in a flask, add enough artificial gastric juice (pepsin 6 g, 35-37% HCl 7 to 10 ml and distilled water upto1000 ml) to more than cover the contents and incubate at room temperature for several hours. Transfer some of the sediment to a slide and examine under a microscope. Further, after removing the digestive juice, the material is put in Baermann's apparatus for collection of larvae.

Fixation of parasites

Fixation means killing a parasite (or any other organism) in a manner that it retains approximately the same shape and size (or without any distortion of its tissues), as it had been while alive. This is generally achieved by use of chemicals. A selective list of the fixatives recommended for helminths is given below:

- 10% formalin (for trematodes, cestodes and nematodes)
- 70% alcohol (for trematodes, cestodes and nematodes)
- Corrosive sublimate-acetic acid fixative (for trematodes and cestodes)
- Alcohol-Formol Acetic Acid (AFA) fixative (formalin commercial 8 parts; alcohol (95%) 50 parts; acetic acid glacial 2 parts; distilled water 40 parts)
- Bouin's fluid (picric acid saturated aqueous solution 75 parts; formalin commercial 20 parts, glacial acetic acid 5 parts; buffer solutions)
- Zenker's fluid (sodium sulphate 1 g; potassium dichromate 2.5 g; mercuric chloride 5 g; distilled water 100 ml; glacial acetic acid 5 ml)

Anyone of these fixatives can be used depending on its availability and/or the preference of the person using it.

Staining of parasites

The purpose of staining is to stain the internal organs of a helminth parasite in a manner that its features and topography become easily visible. However, staining is applied for trematodes and cestodes only. Depending on the availability anyone of the following staining solutions can be used:

- Borax carmine
- Acetic- alum-carmine
- Semichon's carmine
- Harris' hematoxylin

It is necessary to remove all traces of the fixative completely before placing the specimens of trematodes or segments of cestodes in the staining solution otherwise, it will interfere with staining. This is done by washing the specimens in running tap water for several hours and is best done by placing the large specimens (also segments of tapeworms) in a wide-mouthed bottle and small specimens in a Petri dish, and tying their mouths securely with a muslin cloth. These are then placed under continuously running tap-water for about 12 hours (make sure that the water from the tap is not gussing out but flowing slowly). Once the removal of the fixative has been assured, the following steps are to be taken:

- Transfer the specimens to a Petri-dish of a desired size (this will vary according to the size of specimens)

- Treat the specimen for 15 minutes to 1 hour (depending on the size) each successively with 30%, 50% and 70 % alcohol. It is preferable to change the alcohol from the Petri-dish with a dropper rather than removing the specimens from the grade of alcohol to another
- Allow them to remain in 70 % alcohol until stained

Staining with alcoholic solution of borax carmine or acetic-alum-carmine or Semichon's carmine

- Submerge the specimens in a solution of stain (equal volume of stain mixed with equal volume of 70% alcohol) for several hours so as to over stain the specimens
- Withdraw the stain and remove excess by washing twice in 70% alcohol
- Destain in acid alcohol (0.5% to 1% HCl in 70% alcohol) until the specimens are very light pink in colour. However, it is advisable to monitor the destaining process under a dissecting microscope from time to time. The process of destaining can be accelerated by increasing the strength of acid in 70% alcohol
- Once the desired destaining has been achieved, wash the specimens in 70% alcohol 3 to 4 times (by changing the alcohol)
- Dehydrate the specimens by placing them successively in 80% and 95% alcohol for 1 to several hours each. To ensure complete dehydration, it is recommended to make a second change in 95% alcohol or treat with absolute alcohol
- Clear the specimens by placing in a Petri-dish or watch-glass containing either cedar wood-oil or clove-oil. Specimens, when clear, will sink at the bottom of the clearing solution which may take an hour or more
- Give a quick wash in xylol (optional) and mount the specimens in Canada Balsam (which should not be too thick or too thin). Consistency of Canada Balsam can be modified by adding xylol to it

Staining with aqueous solution of borax carmine or acetic-alum-carmine or Semichon's carmine

- Transfer the specimens after removal of the fixative to a Petri-dish containing distilled water
- Add stain to water to get desired intensity of colour (stain to be diluted 1:9 in water)

- Allow the specimens to remain in stain for several hours (preferably over night) to over stain the specimens
- Stain is replaced by 2 or 3 changes of water
- Destain in acid-water (0.5% to 1.0% HCl or 5% HCl if the specimens are thick and muscular) until the specimens are light pink in colour (destaining to be monitored accordingly)
- Once the desired destaining has been achieved, wash the specimens 3 to 4 times, in water
- Dehydrate the specimens by treating them successively with 30%, 50% and 70% alcohol for 1 hour each and with 90% and absolute alcohol for several hours each
- Clear the specimens in either cedar-wood or clove-oil and mount the specimens in Canada Balsam in the same way as mentioned above

Clearing of nematodes

Nematodes are generally not stained and are examined directly by clearing them with a clearing agent. The clearing agents are used according to the size of the nematodes and are as follows:

Glycerine alcohol: It is used for very small nematodes.

Lactophenol: It is the most preferred clearing agent and used as follows:

- Specimens fixed in 10% formalin should be given quick wash 2 to 3 times in normal saline and treated successively with 30%, 50% and 70% alcohol, for 10 minutes each
- If the specimens have been fixed in 70% alcohol, they do not need the above treatment
- Pour lactophenol over specimens in a beaker or a Petri-dish to completely cover them. Also cover the mouth of the container
- Allow it to rest till (may take 1 hour or more) the specimens become transparent and mounted done with the help of DPX and Canada balsam.
- If specimens are large or relatively large, they can be examined in the Petri-dish itself under a dissecting microscope. Specimens can also be manipulated with a needle to study their morphological features

ARTHROPODS

Among the ectoparasites the most commonly encountered in domestic animals are lice, fleas and ticks. These are relatively large arthropods and are hence easy to collect.

Collection of ectoparasites (from the body of the host)

- Dust the animal (host) with an insecticide to immobilize or kill the ectoparasites
- Comb the parasites out of the coat or feathers on to a dark, smooth paper
- Engorged, attached nymphs and/or adult ticks should be removed from the skin of the host with forceps
- Transfer the collected parasites to a bottle of 70% alcohol (or 10 % formalin) which serves as a preservative

Immediate examination of ectoparasites

- Transfer the specimens collected directly from the body of the animal or preserved in 70% alcohol to a test-tube containing a small amount of 10% KOH
- Heat gently for few minutes
- Remove the specimens from the tube and place them on a microscopic slide
- Cover them with a drop of lactophenol and put a cover-slip
- Allow the slide to stand for several hours (24 hours) so that the lactophenol can clear the specimens
- Examine under a microscope under low power magnification

Permanent mounts of ectoparasites

- Place a few specimens taken out from 70% alcohol on a microscopic slide
- Cover specimens with a drop of Hoyer's solution and put a cover- slip and allow it to dry
- Examine under a microscope

Remark: Hoyer's solution serves dual purpose. It fixes and clears the specimens for identification and makes a permanent mount of the parasites.

Composition of Hoyer's Solution:

Distilled Water	50 ml
Crystalline gum arabic (or Gum acacia)	30g
Chloral hydrate	200g
Glycerin	20 ml

Permanent preparation of tick, lice and flea are made by processing in 10% KOH or NaOH, overnight washing, dehydration in ascending grades of alcohol, clearing in clove oil, dipping in xylene and mounting in DPX or Canada Balsam.

Collection and preservation of dipteran flies

The flying insects visiting animals are caught with the help of an insect net. The flies are put in the cyanide killing jar/bottle, the bottom of which has a paste of potassium cyanide and Plaster of Paris about 1.5 cm thick cover with 2 to 3 layers of filter paper. Few paper strips are also placed in the bottle to prevent injury to mouth parts and appendages. The killing can also be done in bottles/ jars containing cotton swab soaked in chloroform. The insects can be preserved in 70% alcohol alongwith few drops of glycerine. Killed flies are pinned vertically by means of fine stainless steel thin pins between prothorax and mesothorax on a cork sheet. These are stored in wooden boxes containing naphthalene balls to prevent fungal growth. Labels are also pinned with the insects. Permanent preparation of mouth parts, legs, wings and posterior spiracles of the larvae are made by dehydrating the material in ascending grades of alcohol, clearing in xylene and then mounting in DPX or Canada Balsam.

6

Immuno-molecular Techniques for Diagnosis of Parasitic Infections

Diagnosis of parasitic infections is routinely done with conventional parasitological techniques as described earlier. However, when the parasitaemia is too low to be detected by the conventional methods as in subclinical infections and/or carrier animals advanced diagnostic tools are used for the early, sensitive as well as specific diagnosis.

Serological techniques

Diagnostic techniques used mostly in laboratories needs various reagents and equipments to perform the test. However, a diagnostic kit is the extrapolation of the same method by packing it in a user friendly format, accompanied by all or most of the reagents required to perform the test and can also be done outside the lab or under field conditions. Various immunodiagnostic kits are available which are based on different diagnostic methods like Enzyme-linked immuno sorbent assay (ELISA). Complement fixation test (CFT), Agar gel immunodiffusion test (AGID), Immunofluorescent tests (IFT), Latex agglutination tests (LAT) etc. ELISA is the most commonly used test system and numerous kits based on ELISA are available commercially.

Agglutination: Interaction between antibody and a particulate antigen results in visible clumping called agglutination. The antigen can be fixed onto inert particles such as latex or bentonite and made to react with serum to produce agglutination. This is called Latex particle agglutination. This test can be read by naked eye. However, these tests suffer from the inherited disadvantages of low sensitivity and specificity. Card Agglutination Trypanosomal Test (CATT), SURATEX are commercially available and are used for diagnosis of trypanosomosis. Modified Agglutination Test (MAT) kits are currently being used for diagnosis of toxoplasmosis.

Agar gel immunodiffusion test: Precipitation reactions can take place in semisolid media such as agar gels. When soluble antigen and antibodies are placed in previously punched out wells in a gel, they diffuse towards each other and where they meet at or near optimal proportions and a precipitin line is formed in the gel and this is called immunodiffusion. Unknown antigens or antibodies can be identified using this technique and can also be employed to show relationship between the antigens placed in adjacent wells. The test is less sensitive and non-specific. It is mostly employed for primary screening of experimental and hyperimmune sera raised for immunological studies.

Complement fixation test: This test uses sheep RBCs sensitized with rabbit antibody as an indicator system. In the presence of complement (usually obtained from guinea pig serum), the sensitized cells are lysed. The test serum is heated at 56°C for 30 minutes to destroy complement in the sample before titration. Antigen is then added to the titrated serum and a precise amount of guinea pig serum complement is also added. After incubation at 37°C for 30 minutes, sensitized sheep RBCs are added followed by a further incubation. Where the complement is fixed by antibody in the test serum reacting with specific antigen, the sheep RBCs do not lyse, but the sensitized sheep RBCs will lyse if complement was not fixed because the test serum did not contain specific antibody. A kit has been developed in India by National Research Centre on Equines, Hisar by the name of COFEB-KIT for detection of equine babesiosis, an economically important disease of horses, ponies, donkeys and mules caused by *Babesia equi.*

Immunofluorescence test: In case of direct immunofluorescence, conjugated antibody is added directly to a smear or tissue section fixed on a slide and after incubation the unbound antibody is removed by washing. The slide is examined under a fluorescence microscope. Where the labeled antibody binds to antigen, bright fluorescence is evident. In case of Indirect fluorescent antibody test (IFAT), the test serum is added to a known antigen fixed on a slide and following incubation and washing, FITC-labelled antiglobulin is added. After washing, the slide is examined for fluorescence. The indirect test is more sensitive than the direct test as more labeled antibody is attached per antigenic site in it. IFAT is widely used for diagnosis of canine ehrlichiosis, babesiosis, toxoplasmosis, theileriosis and trypanosomosis in domesticated animals.

Enzyme immunoassays: ELISA can be designed to detect antigens or antibodies. It is highly sensitive and less than 1 ng of antigen per ml can be detected in specimens. Alkaline phosphatase and Horse radish peroxidase are the enzymes commonly used in ELISA. The colourless substrate is converted to coloured product by the enzymes. ELISA is the serological assay of choice for the qualitative or quantitative determination of antibodies. The advantage in ELISA is that quantitative assays can be based on a single dilution of serum. The wells of the microtitre plate are coated with purified antigen. Types of ELISA:

Direct ELISA: In this test, soluble antigens/antibodies from the specimen are allowed to bind to the captured antibodies/antigens. After the unbound components are washed away, an enzyme labeled antibody is added. After washing, the substrate for the enzyme is added. The readout is based on the colour change that follows.

Indirect ELISA: In this test, the antigen is attached to solid phase (e.g., the wells in polystyrene plates or nitrocellulose membrane). The antiserum is reacted with this immobilized, specific antigen and the specific antibody present will bind to the antigen. Exposure of this complex to enzyme labeled anti-immunoglobulin antibody results in binding to any specific antibody molecules absorbed from the serum. The complex is washed and the substrate for the enzyme is added resulting in activity proportional to the amount of specific antibody in the original serum. Indirect ELISAs have greater sensitivity and the need to label each specific antibody is avoided.

Labelled antigen ELISA: The labelled antigen is bound to the plate before testing. The antibody to be tested is added followed after washing. The antibody binds the labeled antigen to the plate.

Avidin-biotin system: In this, the antibody is conjugated to biotin. The biotin has extraordinary high binding affinity for avidin. It does not alter the antigen binding capacity of antibody. The antigen-antibody complex is recognized with high sensitivity by adding Avidin-labeled enzyme and then the substrate. ELISA may be used for assaying antigens by either a competitive or a double antibody method and for assaying a specific antibody by an indirect method.

Competitive ELISA: In this, a mixture of a known amount of enzyme labelled antigen and an unknown amount of unlabelled antigen is allowed to react with specific antibody attached to a solid phase. After the complex has been washed with buffer, the enzyme substrate is added and the enzyme activity is measured. The difference between this value and that of a sample lacking unlabeled antigen is a measure of the concentration of the unlabelled antigen.

Double antibody (Sandwich) ELISA: In this test, the unknown antigen solution is reacted with specific antibody attached to a solid phase, washed and treated with enzyme labeled antibody. After a further wash, the enzyme substrate is added. The amount of enzyme activity measured under standard conditions is directly proportional to the amount of antigen present.

Dot-ELISA: In this test, nitrocellulose membrane is used to immobilize the antigen and the antigen is placed on the nitrocellulose membrane as a spot. The development of a brown spot is the positive reaction in this test.

Slide-ELISA: In this test, microscopic glass slide is used instead of nitrocellulose membrane to immobilize the antigen.

Note

1. ELISA has been widely used for the diagnosis of parasitic diseases. Previously, ELISA based on whole cell lysate/antigen was used but it suffers from the problem of cross-reactivity. Therefore, to overcome this problem, nowadays, recombinant protein antigen based ELISA is being employed. Important parasitic diseases like amphistomosis, fasciolosis, toxoplasmosis, trypanosomosis, cysticercosis, haemonchosis and babesiosis have been diagnosed using various types of ELISA.
2. A commercial kit based on antigen detection of *Cryptosporidium parvum* oocysts in animals (Fig 9) and antibodies against *Ehrlichia canis* in dogs utilizing serum, plasma or blood (Fig 10) is available in the market. Further, a test kit for diagnosis of *Neospora caninum, Babesia bigemina, Anaplasma marginale* and *Giardia* antibodies is also commercially available.

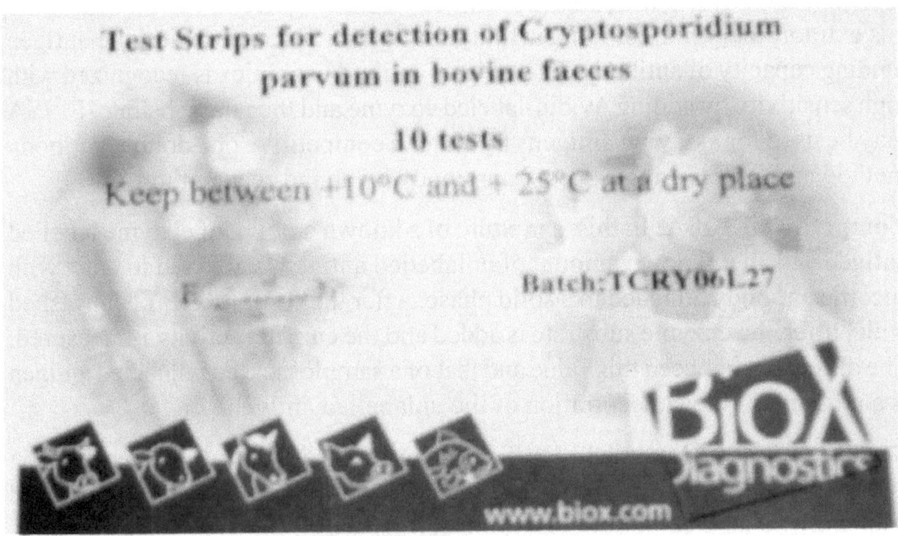

Fig. 9: Kit for diagnosis of *C. parvum*

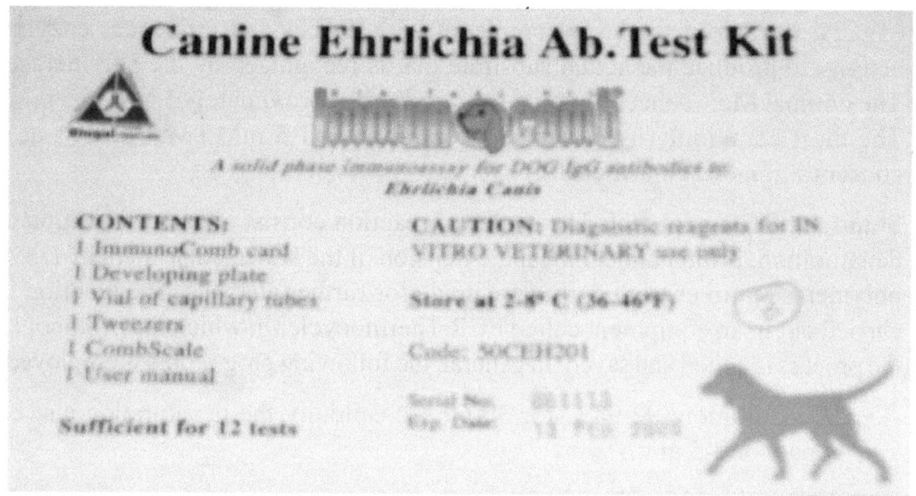

Fig. 10: Kit for diagnosis of *E canis*

Molecular diagnosis

Polymerase Chain Reaction (PCR), developed by Kary B. Mullis in 1984, is a technique to exponentially amplify *in vitro* a small quantity of specific nucleotide sequence present in the template of an organism by using specific set of small oligonucleotide sequences called as primers. It is highly sensitive test and can detect minute quantity of organisms.

PCR reaction components

Template: The amount of template in a reaction strongly influences performance in PCR. The recommended amount of parasitic DNA for standard PCR is 1–10 ng.

Primers: These are small oligonucleotide sequences (preforably 18-24 bp long) that hybridize to opposite strands and flank the region of specific nucleotide sequence in the target DNA. The primers are of two types: forward and reverse. Optimal primer concentrations between 0.1 and 0.5 µM are generally used in PCR reaction. For most application 0.2 µM produces satisfactory results.

Thermostable DNA polymerase: For most of the PCR protocols used in routine Taq DNA polymerase is the standard PCR enzyme and the optimum concentration of enzyme is between 0.5 and 2.5 units/50 µl reaction volume.

Deoxynucleotide triphosphate (dNTP): The final dNTP concentration should be 50–500 µM (each dNTP). The most commonly used dNTP concentration is 200 µM.

$MgCl_2$: Mg^{2+} forms soluble complexes with dNTPs and influences enzyme activity to produce the actual substrate that is recognized by the polymerase. The optimal Mg^{2+} concentration may vary from approximately 1 mM to 5 mM. The most commonly used Mg^{2+} concentration is 1.5 mM (with dNTPs at a concentration of 200 μM each).

Standard PCR protocol: The cycles of reaction consist of repeated template denaturation, primer annealing and extension of the annealed primers by DNA polymerase until enough copies are made for further analysis. The reaction is carried out in an equipment called PCR Thermocycler in which the protocol of the process is edited and saved. In general, the following programme is employed:

- Denaturation: 92-94°C for 30-60 sec (initially the denaturation can be done for 2-5 min.)
- Annealing: 37 - 72°C for 30-60 sec
- Elongation: 72°C for 30-60 sec

Repeat these steps for 25-32 cycles only (otherwise enzyme decay will cause artifacts)

- Final elongation/extension: 72°C for 5 min at end to allow complete elongation of the product.

Visualization of the product: The products are checked for amplification by agarose gel electrophoresis on an agarose containing ethidium bromide using 50 V for 1 h. The gel is then visualized using gel documentation system and may be photographed for record.

Note

1. Different types of PCRs have been used employing whole blood/ blood clots/tissues/secretions and fluids for the diagnosis of various parasitic diseases like babesiosis, toxoplasmosis, trypanosomosis, ehrlichiosis, hepatozoonosis, theileriosis and anaplasmosis.
2. The PCR assays have the advantage of higher sensitivity and specificity and hence can be used to detect the carrier animals or chronic cases of infections. The carrier animals are an important source of infection to the susceptible animals hence, their identification is of primary importance in order to formulate effective control strategies against these diseases.
3. PCR assay like multiplex PCR has been used for simultaneous diagnosis of multiple infections in animals e.g. a multiplex PCR assay has been developed for detection of *Babesia*, *Theileria* and *Anaplasma* infection in cattle.

4. Similarly, to differentiate between two parasitic species PCR- RFLP method has been developed e.g. PCR-RFLP method has been developed employing apicomplexan taxon specific primers and subsequently restriction enzyme analysis to differentiate between *Babesia* and *Theileria* infections. On similar lines, these parasites have been differentiated on the basis of B-tubulin gene fragment.

7

Chemotherapy Against Parasites

Chemotherapy in Parasitology deals with the treatment of disease caused by different parasites by specific drugs; meanwhile particular drug has got the particular action against the particular parasite. Recently chemotherapy is a extensively used and emerging branch for the control of different stages of parasites that is normally not killed by the conventional methods of treatment. Thus, chemotherapy plays an important role in the treatment and control of parasitic infections by avoiding the outcome of the danger of the disease by treatment of the affected animals. They should have the maximum parasitotrophic action and minimum organotrophic action. It is mainly divided into two types:

1) Rational therapy: It is a mode of treatment based on the pharmacological actions of drugs with relation to disease.
2) Supportive therapy: It is a method of treatment other than drugs, e.g. giving high protein diet in the oedematous condition such as bottle jaw in fascioliosis and amphistomiosis.

The drugs used for the killing or complete removal of the parasites from the body of the host called antiparasitic drugs.

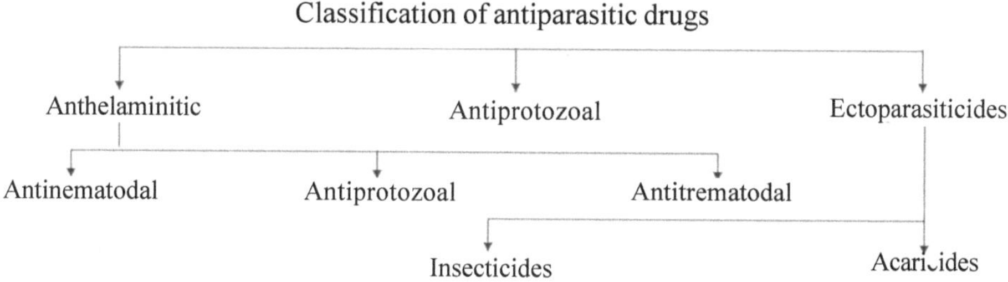

Anthelmintic

The drugs which are used for the treatment of helminth parasites are called **anthelmintic**. These are the widely used drugs for the treatment and control of parasitic infection in the veterinary practices. Drugs used for the killing of worms within the body cavity are known as **vermicides** and the drugs used for the expulsion of the parasites from the body of the host are known as

vermifuges. Some of the anthelmintic drugs in lower dose of concentration act as a vermifuge and higher dose of concentration acts as vermicidal. The drugs having activity against both endoparasites and ectoparasites are called **endectocide** such as ivermectin, closantel etc.

Properties of ideal anthelmintic drugs

1) Should have broad spectrum of activity
2) Should be effective against both immature/larval and mature/adult stages of the parasites
3) Should have wide margin of drug safety index and high therapeutic ratio
4) Should not interfere with normal physiological functions of the body of the host
5) Should have least residual period in the tissue or milk i.e. minimum with holding period
6) Should be easy to administered i.e. preferably orally in a single dose or in divided dose in the same days
7) Should be economical and easily available in the market and stable for long period
8) Should be compatible with feed or other compounds
9) Drugs should reach that portion of the body where the worm infection occurs
10) Drugs must not cause irritation to GI tracts and should not produce nausea, diarrhea or vomiting
11) Drugs should not be absorbed from the GI tracts when used for intestinal helminthes while they should be absorbed in case of tissue helminthes
12) Should have high therapeutic ratio: The ratio between maximum tolerated dose and minimum curative dose

Classification of antinematodal drugs

1. Benzimidazole compounds

   Thiabendazole, Albendazole, Fenbendazole, Mebendazole, Oxibendazole, Oxfendazole, Flubendazole, Parbendazole, Cambendazole etc.

   Pro-benzimidazole drugs: Prodrug of Fenbendazole- Fabantel

   Prodrug of Albendazole- Netobimin

   Prodrug of Lobendazole- Thiophanate

2. Avermectins and Milbemycins or Macro cyclic lactones:
   a) Avermectins - Ivermectin, Doramectin, Abamectin, Salamectin, Eprinomectin
   b) Milbemycins - Milbemycin oxime, Moxidectin, Nemadectin
3. Imidazothiazoles - Tetramisole, Levamisole, Butamisole hcl
4. Tetrahydropyrimidines - Pyrantel, Morantel, Oxantel
5. Organophosphorus compound (OPC) - Dichlorovos, Trichorfon, Coumaphos, Haloxon, Crufomate, Naphthalophos
6. Arsenicals - Thiacetarsamide, Melarsoamine, Arsenamide, Glycobiursol
7. Substituted phenol and salicylanilides - Disophenol, Nitroscanate, Nitroxynil, Closantel
8. Simple heterocyclic drugs - Phenothiazine, Piperazine, Diethylcarbamazine citrate (DEC)
9. Miscellaneous drugs - Tetrachloroethylene, N-butylchloride, Toluene, Thenium, Hygromycin-B, Pthalofyne, etc.

Classification of anticestodal drugs

1. Synthetic organic compounds:-
   a) Isoquinolones: Praziquantel, Epsiprantel
   b) Salicylanilides: Niclosamide
   c) Substituted phenol: Dichlorophen, Bithionol, Nitroscanate
   d) Benzimidazole and pro-benzimidazoles
   e) Miscellaneous: Bunamidine, Resorantel, Uredophos
2. Natural organic compounds: Rhizome of male Fern, Kamala, Nicotine sulfate and Arecholine
3. Inorganic compounds: Lead arsenate, Tin oxide and Tin chloride

Classification of antitrematodal drugs

1. Substituted phenol: Nitroxynil, Hexachlorophen, Trichlorophen, Bithionol, Niclofolan
2. Salicylanilides: Closantel, Rafoxanide, Oxyclozanide, Niclosamide, Brotianide

3. Aromatic amides: Diamphenethide (immature flukes)
4. Sulfonamides: Clorsulon
5. Halogenated hydrocarbans: Carbon tetrachloride, Hexachloroethane, Hexachloroparaxylene, Tetrachlorodifluroethane
6. Benzimidazole and pro-benzimidazoles: Triclabendazole, Albendazole
7. Isoquinolones: Praziquantel

The protozoan parasites are highly pathogenic as they are causing harmful and deadly diseases in the livestock populations. Thus, the control of these infections is mainly based up on with antiprotozoal treatment.

Classification of antiprotozoal drugs

1. Anti-coccidial drugs

1) Thiamine antagonist- Amprolium
2) Ionophore/Polyether antibiotic- Monensin, Lasalocid, Salinomycin, Narasin, Maduramycin Semduramycin Laidlomycin
3) Folic acid antagonists- Sulfonamides- Sulphaquinoxaline, Sulphaguanidine, Sulphadiazine, Sulphamethoxine, Sulphadimidines

   Pyrimidines- Trimethoprim, Pyrimethamine, Ormetoprim,

   Substituted benzoic acid- Ethopabate + amprolium combination

   Potentiated sulfonamides- these combinations as given below are very much effective:

   Trimethoprim-sulphadimethoxine

   Trimethoprim-sulphadiazine

   Pyrimethamine-sulphadoxine etc.
4) Quinolones- Decoquinate, Methylbenzoquate, Buquinolate
5) Pyridinoles- Clopidol
6) Guanidines- Robenidine
7) Nitrobenzamide- Dinitolmide, Aklomide
8) Carbanilides- Nicarbazin
9) Quinazolines- Halofuginone
10) Benzene acetonitrile- Diclazuril, Clazuril

The mechanism of action, spectrum and dosage regimen of anthelmintic drugs

Drugs	Mechanism of action	Anthelmintic spectrum	Dosage regimen
Benzimidazole	1. Act specifically by binding on β-tubulin dimer and inhibition of polymerization of tubulin into microtubule formation leads to disrupt integrity and transport function of the cell within the parasite and block cell division. 2. Inhibition of the mitochondrial fumarate reductase, blocking of glucose transport and uncoupling of oxidative phosphorylation lead to inhibition of essential energy to the parasites by giving in divided dose more appropriately than the single dose.	Broad spectrum activity against all major GI nematodes of cattle, sheep, goats, pigs and horses. These drugs used to treat mostly helminthes like *Ascaris, Toxocara, Strongylus, Oxyuris Trichostrongylus, Taenia, Fasciola* etc. The complete excretion of thia bendazole occurs within 5 days of single oral administration. Albendazole is contraindicated in lactating dairy cattle. The withdrawal period of fenbendazole is 14 days for meat and 72 hrs for milk. Mebendazole and cambendazole is not recommended in pregnant animals because it has shown to be embryotoxic and teratogenic in experimental animals. The peak plasma concentration of fabantel occur in 6-12 hrs after dosing in sheep and 12-24 hrs in cattle. The withdrawal period for triclabendzole is 28 days.	Albendazole: For nematodes & cestodes, Cattle@7.5mg/kg b.w., PO Sheep @5mg/kg b.w., PO Goats@10mg/kg b.w., PO For trematodes, Cattle @10mg/kg b.w., PO Sheep @7.5mg/kg b.w., PO Thiabendazole: Cattle@50-100mg/kg b.w., PO as a single dose Sheep, goats @50-75mg/kg b.w., PO as a single dose Horse @50-110mg/kg b.w., PO as a single dose. It is effective against mature roundworm only. Fenbendazole: Cattle, goat @7.5mg/kg b.w., PO as single dose Sheep, @5mg/kg b.w., PO as single dose Mebendazole: Dog and cat @25-50mg/kg, b.w., PO, bid for 5 days. Sheep @10-15mg/kg b.w., PO as a single dose Triclabendazole: Sheep &goats @10mg/kg b.w. PO, Cattle & horses @12mg/kg b.w. PO against *Fasciola* infection. Cambendazole: Horse, cattle and sheep @20mg/kg b.w., PO Parbendazole: Cattle, sheep and pig@30mg/kg b.w., PO as a single dose.

Contd.

			Flubendazole: Pigs @ 5mg/kg b.w., PO as a single dose Oxfendazole: Cattle, sheep @ 5mg/kg b.w., PO as single dose, horse @10mg/kg b.w., PO as a single dose. Oxibendazole: Horse@10mg/kg b.w., PO as a single dose (15mg/kg for *Strongyloides*), cattle and sheep@10-20mg/kg b.w., PO
Piperazines	1. Neuromuscular blocker anthelmintic. 2. Mechanism of action is similar to ivermectin by blocking action of acetylcholine to produce anticholinergic action at neuromuscular junction and to block succinate production by the worms lead to deplete energy of the parasites followed by flaccid paralysis.	Roundworm (*A. suum*), pin worm (*E.vermicularis*) infection in pig as well as human being. *Toxocara vitulorum* in cattle and Parascaris equorum in horse. The excretion of this compounds starts early within 30 minutes and is virtually complete with in 24 hrs.	Piperazine @ 80-100mg/kg b.w. PO for Ascariosis and *Thelazia* spp. For ascarids: Dogs and cats @55-110mg/kg, PO
Macrocyclic lactones	These compounds specifically acts on GABA-chloride ion gated channel of parasite post synaptic membrane lead to increased chloride ion concentration by opening channel followed by membrane hyper polarization lead to flaccid paralysis of parasites.	Widely used for the control of endoparasite and ectoparasite in domestic animals i.e. GI nematodes lungworms in horses, ruminants, pigs, heartworm and mites, mange, lice, grubs etc. Ivermectin has a long terminal half-life (dog, 2 days; cattle, 2-3 days; sheep, 2-7 days; pig,	For lung worm infection: Cattle @0.2mg/kg b.w. i/m for successive 3 days. Diethylcarbamazine (DEC): For heartworm infection: Dogs @ 6.6mg/kg b.w., PO, once daily. Ivermectin @0.2mg/kg b.w. s/c or 200mcg/kg b.w. s/c; avoid in collie bred of dog and young calves and pups; in case of horse given only by orally. Doramectin: cattle, sheep@0.2mg/kg b.w., s/c pig @0.3mg/kg b.w., i/m Abamectin: cattle @0.2mg/kg b.w., s/c Selamectin: dogs and cats@6mg/kg b.w., topi cally; may be repeated after 30 days. *Contd.*

		0.5 days). Avoid use in lactating animals. The withdrawal period for cattle is less than 4-9 days.	Eprinomectin: cattle @0.5mg/kg b.w., topi cally Milbemycin oxime: for prevention of heart worm and treatment of GI nematodes, in dogs and cats @0.5mg/kg b.w., PO, once . But it is not effective against cestodes and trematodes.
Imidazothiazole	Cholinomimetic drugs- nicotinic like action by acting as autonomic ganglionic stimulants and cause the activation of parasympathetic and sympathetic nervous system followed by sustained muscular contraction lead to muscular paralysis.	Primarily used for treatment of the GI nematodes. The complete elimination of the drug from the body usually occurs within 2 days. No residues of levamisole are detected in tissues, blood or urine 7-8 days after treatment. It also has immunostimulant activity.	Levamisole: cattle, sheep@7.5mg/kg b.w. s/c or PO as a single dose Tetramisole: dogs@2mg/kg b.w., s/c Butamisole: dogs@2.2mg/kg b.w., s/c, repeated in 21 days.
Tetrahydropyri-midines	Nicotinic like depolarization neuromuscular blocking agent, causes activation of nicotinic cholinergic receptor and producing slowly developing contraction and spastic paralysis of the parasites.	Hookworm, pinworm and round worm infestations. Pyrantel is absorbed more rapidly with peak plasma concentration in 2-3 hrs in dogs and pigs after dosing. Morantel is slower than pyrantel in its onset of action, but it is several times more potent than pyrantel.	Pyrantel pamoate @14.4mg/kg b.w. in dog and cats. Pyrantel tartrate @25mg/kg b.w. in cattle, sheep and goats. Morantel: cattle@8.8mg/kg b.w., PO sheep @10mg/kg b.w., PO Oxantel: dogs@55mg/kg b.w., PO
Isoquinolones	Initiate rapid muscular contraction by altering ionic balance, Ca++ channel lead to hyperpolarization and spastic paralysis and vacuolation and disruption of its teguments.	Visceral schistosomosis, *F. buski*, Clonorchiosis and Opisthorchiosis, *Dicrocoelium* spp. and cestodes infection. Praziquantel has elimination half-life of approx. 3 hrs in dogs.	In visceral schistosomosis @40-50mg/kg b.w. Praziquantel @5-7.5mg/kg b.w. oral or i/m in dog and cats. It is thought to be drug of choice against cestodes infection and also used against trematode infections
Salicylanilides	Mitochondrial oxidative phosphorylation uncoupler.	*T. saginata, H.nana, D.latum, T. solium* infestations and also effective against *Fasciola*, amphistomes, and GI nematodes like *H. contortus*.	Niclosamide @50-100mg/kg, b.w. orally in cattle, sheep, goats and horses Closantel @5mg/kg b.w. s/c in sheep, goats and cattle or 10mg/kg b.w. PO

Niclosamide rapidly excreted from the body unlike other salicylamilides. Nitroxynil is slowly eliminated from the body, also excreted in milk and is retained in liver and plasma of sheep at detectable level of about 2 months after single dose. Closantel is slowly excreted in faeces (80%) and also urine. It has plasma half-life of about 15 days in sheep. Rafoxanide has long withdrawal period of 28 days, so it is not used in lactating cows. The withdrawal time of oxyclozanide is 14 days.	Rafoxanide @7.5mg/kg b.w. PO and 3mg/kg b.w. s/c, Oxyclozanide @10-15mg/kg b.w. PO

11) Triazinones- Toltrazuril

12) Benzyl purines- Arpinocid

13) Miscellaneous- Oxytetracycline, Chlortetracycline, Furazolidone, Nitrofurazone

2. Antipiroplasmal drugs

1) Diamidine derivatives:

   A) Carbanilides diamidines- Imidocarb, Amicarbalide

   B) Aromatic diamidines- Diminazene aceturate, Pentamidine, Phenamidine isothianate.

2) Quinuronium derivatives- Quinuronium sulphate (Acaprin, Babesan, Pirevan, Piroparv, Piroplasmin)

3) Naphthoquinones- Buparvaquone, Parvaquone

4) Tetracycline- Oxytetracycline, Chlortetracycline, Doxycycline

5) Quinazolines- Halofuginone

6) Dyes- Trypan blue

3. Antitrypanosomal drugs

1) Aromatic diamidines- Diminazene, Phenamidine, Propamidine, Pentamidine, Stilbamidine

2) Quinapyramine compounds- Antrycide Prosalt, Quinapyramine sulphate, Quinapyramine chloride

3) Aminophenanthridine compounds- Homidium, Isometamidium, Pyrithidium, Dimidium

4) Carbamide derivatives- Suramin

5) Antimonial compounds- Antimonial potassium tartrate, Stibophen

6) Dyes- Trypan blue, Trypan red

4. Antigiardial drugs/ Antiamoebiosis drugs/ Antitrichomoniosis drugs

1) Nitroimidazoles- Metronidazole, Dimetronidazole, Tinidazole, Ornidazole, Sacnidazole

2) Benzimidazole- Albendazole, Fenbendazole

3) Nitrofurans- Furazolidone

4) Acridine derivatives-Quinacrine

5. Antihistomonad drugs

1) Nitrothiazoles- Aminonitrothiazole, Nithiazide

2) Nitroimidazole- Dimetronidazole

6. Antitoxoplasmosis drugs- Sulfonamides + pyrimethamine combinations, Clindamycin, Spiromycin

7. Antimalarial drugs- Quinine, doxycycline, tetracycline, pyrimethamine sulfadoxine, artemisinin, chloroquine, amodiaquine, piperaquine, mefloquine, primaquine, tafenoquine

Ectoparasiticides

The drugs used for the prevention and control of ectoparasitic infestations are known as **ectoparasiticides** or which are used for the treatment and control of insects are known as **insecticides** and those against ticks and mites are known as **acaricides**. Therefore, these drugs used for the treatment and control of insects as well as tick and mite infestations.

Properties of an ideal ectoparasiticides

1) Broad spectrum activity, destroy all types of external parasites

2) Effective in all stages of life cycle of the parasites

3) Non-toxic or less toxic to the host animals

4) Stable and longer residual activity on hair and skin surfaces

5) Rapid onset of action

6) Compatible with vehicle or bases required for application

7) Cheap and easily available

8) Drugs have no adverse affects on skin, hair and wool

9) Provide minimum residue in or on the host and be rapidly detoxified in the environment

Drugs	Mechanism of action and its spectrum	Dosage regimen
Amprolium	Act on early first generation schizonts, preventing differentiation of merozoites and also suppress sexual stage and sporulation of oocysts.	Poultry @0.012-0.024%, PO with water for 3 days, Layer: Amprolium+Ethopabate (125ppm +8 ppm) combination used in feed
Monensin	It has affinity for cations and form lipophilic complex ↓ Ionophore-cation complex renders cell membrane permeability to cations allowing excessive entry inside cell ↓ Increase intracellular cation concentration interfere with certain mitochondrial functions i.e. substrate oxidation and ATP hydrolysis in susceptible coccidian parasites ↓ Increase propionic acid production and better FCR leads to decrease CO_2 and methane production. The pre-slaughter withdrawal period is 72 hrs in chicken but have no withdrawal period in goats. The pre-slaughter withdrawal period of sulphaquinoxaline from broiler birds is 5 days. Pre-slaughter withdrawal period of decoquinate is 3 days in broiler.	Broiler @100-200 g/tone feed in poultry or 100-125 ppm in feed. For bovine coccidiosis: Amprolium, Monensin, Decoquinate, Diiodohydroxyquinoline, Sulphamethazine, Sulphamerazine, Sulphaquinoxaline For ovine and ceprine coccidiosis: Sulfaguanidine, Nitrofurazone, Amprolium, Monensin, Sulphaquinoxaline For swine coccidiosis: Amprolium, Sulfadimethoxine
Imidocarb diproprionate (Carbanilamide)	Acts directly by combining with DNA in susceptible protozoa, causing DNA to unwind and denature. It also interferes with DNA synthesis. Most effective therapeutic and prophylactis drugs against *B.bigemina* and *B. bovis*. Imidocarb gives antiprotozoal activity for several weeks (>4 weeks).	Cattle and sheep @1.2mg/kg b.wt. s/c single dose Dog @6.6mg/kg b.wt. i/m, s/c Horse @2.4mg/kg b.wt. i/m, s/c
Diampron (Amicarbalide)	Used as prophylactic drug because of depot preparation i.e. slow release and also for therapentic purpose	@10mg/kg b.wt. by deep i/m
Diminazene aceturate)	Bind to DNA of protozoan parasites lead to the effective killing of the organisms. Diminazene is eliminated slowly from the body.	Dose: @3-8mg/kg b.wt. by deep i/m inj. or 0.8-1.6g/100kg b.wt. Cattle, Horse @2-5mg/kg b.wt. deep i/m

Contd.

	The drug is well tolerated and safe. It is used to treat all species of *Babesia* Used for both *Babesia* and *Trypanosoma* infections	Dog @3.5-7mg/kg b.w. s/c
Buparvaquone	Very effective and highly specific for the treatment of experimental as well as clinical cases at early as well as late stages of *Theileria* infection. Buparvaquone has long plasma half-life of at least 7 days.	Dose: @2.5mg/kg b.w. i/m single dose
Antrycide Prosalt	Inhibition of cell division and growth of the parasites. It is the combination of Quinapyramine sulphate and quinapyramine chloride in 3:2 ratios. These ratio of compounds in which Quinapyramine sulphate act as curative while quinapyramine chloride used for prophylactic treatment of animals. Quinuronium sulphate compounds are generally not repeated for at least a period of 2 weeks due to occasional development of sensitization.	Dose: Quinapyramine sulphate @5 mg/kg b.w. as 10% aq.soln Antrycide Prosalt @7.4mg/kg of a solution of 3.5 g powder in 15 ml of water For treatment of babesiosis: Cattle, sheep and pigs @0.5mg/kg, s/c, Horses@0.3mg/kg b.w., s/c Dogs@0.25mg/kg b.w., s/c
Isometamidium chloride	It is effective against *T. brucei, T. congolense, T. vivax* and provided protection of 3-6 months	Dose: Cattle and horses @0.5-1mg/kg b.w. deep i/m single dose
Suramin	Suramine bind plasma protein to form drug-protein complex which enter the trypanosomes by endocytosis ↓ Complex lysed by the lysosomal enzymes which breaks the proteins ↓ Free suramin act on trypanosomal enzymes particularly trypanosomal cytosolic serine oligopeptidase. It is the drug of choice of surra causing by *T.evansi* organism.	Dose: Horse @7-10 mg/kg b.w. slow i/v Cattle @12mg/kg b.w. slow i/v
Metronidazole	It enters the susceptible protozoa by diffusion and is reduced to intermediate metabolites which interact with the protozoal DNA, where it causes a loss of helical structure and strand breakage. Drug is mainly used for giardiosis and trichomoniosis. The elimination half-lives of metronidazole in animals are 4-5 hrs in dogs and 3-4.5 hrs in horses.	For giardiosis and trichomoniosis:Dog and cats @20mg/kg b.w. PO, OD, for 5-8 days For amoebiosis in dog and cats @ 60mg/kg b.w. OD for 5 days. It is used for hexamitosis and histomoniosis @27g/100 L in drinking water for 12 days
Aminonitrothiazole		It is used in Turkey @0.1% in feed for 14 days.
Nitrothiazole		It is also used in Turkey @0.4% in feed for 7 days.

Method of application of ectoparasiticides

1) Dips
2) Sprays
3) Pour-on
4) Spot-on
5) Ear tag
6) Back rubbing
7) Dusting
8) Collar applied
9) Shampoo
10) Washing
11) Parentally
12) Orally
13) Repellents

Classification of ectoparasiticides

1. Insecticides

1) Pyrethrin and pyrethroid derivatives:
    a) Natural- Pyrethrin
    b) Synthetic- Cypermethrin, Permethrin, Flumethrin, Deltamethrin (Butox), Allethrin, Fenvalerate, Cyfluthrin, Cyhalothrin etc.
2) Organophosphorus insecticide/compounds (OPC)- Dichlorovos, Malathion, Parathion, Coumaphos, Trichlorfon, Chlorpyriphos, Diazinon, Ethion, Fenthion, Chlorfenvinphos, Famphur, Nuvan, Neguvon, Tabin, Sarin, Cythion, Crotoxyphos, crufomate, Dichlorfenthion, Iodofenphos, Phosmet, Ronnel, Temophos
3) Organochlorines compounds- DDT, BHC/Lindane/HCH, Methoxychlor, Chlordane, Endosulphan, Aldrin, Dieldrin, Bromocyclen, Toxaphene
4) Carbamates derivatives- Carbaryl (0.5%), Propoxur (0.5%), Butacarb, Carbanolate
5) Formamidines derivatives- Amitraz
6) Chloronicotinyl- Imidaclaprid
7) Phenylpyrazole- Fipronil
8) Botanical-Rotenone, Nicotine

2. Avermectins & Milbemycins/ Macrocyclic lactones

A) Avermectins- Ivermectin, Doramectin, Abamectin, Salamectin,

B) Milbemycins- Milbemycin oxime, Moxidectin

3. Insect growth regulator

A) Juvenile harmone analogue- Methoprene, Cyromazine, Pyriproxyfen, Fenoxycarb

B) Chitin synthesis inhibitor- Lufenuron, Diflubenzuron

4. Miscellaneous drugs- Benzyl benzoate, Sulphur and Borates

Insecticides	Mechanism of action and its spectrum	Dosage regimen
Pyrethrins	It is absorbed through sensory organ and body of the parasite ↓ Nerve signal abolished by keeping sodium channel open ↓ Causing marked over excitation with tremor and spasms ↓ Paralysis in the body of parasite followed by death. There is little or no accumulation of pyrethroids in the body tissues.	Dose: Mostly used in concentration of 0.1-0.9% in combination with 0.5-1% piperonyl butoxide in sprays, foam, dust, shampoo, dip, ear tag against flea, lice, tick, mites in dog and cat and sprays against flea and lice in horses.
OPC and Carbamates	Acts both as stomach and contact poisons ↓ Reversible inhibitor of AchE enzyme (carbamates) and irreversible inhibitor of Ach E enzyme (OPC) ↓ Both leading to the spastic paralysis of the parasites OPC are mostly rapidly eliminated from the body; so their tissue or blood residues are minimal within few hrs of exposure. Carbamates are rapidly eliminated mainly as metabolite in urine.	Dose: Malathion in large animal @0.2-0.5% Small animal @ 0.5% solution for topical application in the form of spray/dip Dicholorvos (DDVP) used in flea and tick collar in dogs and 1 % spray for flea, ticks, mites and lice Coumaphos @15mg/kg in cattle
Organo-chlorines	Act as contact poison ↓ They cause CNS stimulation by excess release of Ach at the end of nerve axon ↓ Leading to spastic paralysis of the parasites. The half-lives of some organochlorine insecticides may ranges from days to weeks, residue in body tissue may be found for several months, possibly years	Dose: Lindane as dusting powder 0.2-0.5% Bath 0.2% Dip and spray 0.02 - 0.05%

Amitraz	Monoamine oxidases inhibitor and also inhibits prostaglandins synthesis. The elimination half-live is approx. 24 hrs in dogs. After a spot-on application, its distributes throughout hair coat reaching maximum concentration in 7-14 days and quantifiable levels for up to 56 days.	Dose: Dog @0.025-0.05% solution
Fipronil	Exerts its action by selective blocking the passage of chloride ions through GABA-receptor and glutamate-receptor which disrupts insects CNS. The elimination half-live is 7-8 hrs and fipronil sulphate is 7-8 days.	Dose: 0.75-25 ml of 10% solution

There is wide variety of different anthelmintics, antiprotozoals and ectoparasiticidal drugs are available in the market. But due to improper or excessive use of these chemicals, the problems of anthelmintic resistance, ectoparasiticidal resistance, acaricide resistance were reported worldwide including India.

8

Methods of Detection of Antiparasitic Drug Resistance

A. Evaluation of the efficacy and resistance of anthelmintics against Helminths

Anthelmintic resistance

According to WAAVP (World Association for the Advancement of Veterinary Parasitology) anthelmintic resistance is a failure to reduce feacal nematode egg count by at least 95%. Technically accurate definition is that resistance is a genetically determine decline in the efficacy of an anthelmintic against a population of parasite that is susceptible to the drug. The history of parasite resistance to anthelmintic starts with the first report on phenothiazine resistance in 1957. *Haemonchus contortus* was the first nematode to develop resistance against different anthelmintic.

The World Association for the Advancement of Veterinary Parasitology has recommended two tests for detecting anthelmintic resistance in ruminants, horses and pigs. These are *in vivo* test (FECRT) and an *in vitro* test (the egg hatch test).

Detection Technique

Test	Specturm	Application
In vivo		
1. FECRT	All	Widespread
2. Critical anthelmintic test	Anthelmintic	Limited
3. Controlled slaughter test	Anthelmintic	Limited
In vitro		
1. Egg hatch assay	BZ/LEV/Morantel	Widespread
2. Larval paralysis/motility test	LEV/Morantel	Limited
3. Tubulin binding assay	BZ/All drugs	Limited
4. Larval development assay	Benzimidazoles	Commercialized
5. Adult development	All drugs	Limited
6. Molecular based test (AS-PCR and PCR-RFLP)		Limited

FECRT (Faecal Egg Count Reduction Test)

i. Choose animal with egg count >300 epg.

ii. Use 10 animals per group if possible.

iii. Rectal sample putting 3-5 g into individual pots.

iv. Count using the Mc Master technique as soon as possible after collection

v. Only store at 4^{0}C for 24 h if using sample for culturing.

vi. Individually weight animal and give manufacture recommend dose orally, from a syringe.

vii. Take second rectal sample at the following time after treatment.

viii. Levamisole 3-7 days, Benzimidazole 8-10 day, Macrocyclic lactone 14-17 days.

ix. If testing all groups in same flock collect at 14 days.

x. For interpretation of result

$$\text{FECRT\%} = \left(1 - \frac{X_t}{X_c}\right) \text{x } 100$$

X is anthelmintic mean EPG 10-14 day post treatment

t is the treated and c is the untreated control group.

EGG HATCH ASSAY (EHA)

Collect fresh fecal sample from suspected flock directly from rectum

Separate fresh egg from faeces.

1. Soaked the faeces in the water for 15 minute and prepare faecal slurry.
2. Strain the slurry and centrifuge at 2000 rpm for 5 minutes
3. Discard supernatant and re-suspend the sediment containing egg in saturated salt solution
4. Mix the content well and centrifuge at 2000 rpm for 1 minute
5. Cover test tube with cover slip and allow the egg to float
6. Lift the cover slip gently and collect egg in the beaker containing water by dipping the cover slip in the water

- Estimate number of egg suspension and make final concentration around 150 eggs/ml

- Preparation of TBZ solution : dissolve 10 mg of TBZ in 5 ml DMSO (20 μg TBZ/10 μl) and use it as stock solution (A). Make serial dilution of TBZ solution with distilled water as follows:

Tube no.	1	2	3	4	5
Volume of solution A ml	0.5	-	-	-	-
Volume of solution from tube 1(ml)	-	1.0	-	-	-
Volume of solution from tube 2(ml)	-	-	1.0	-	-
Volume of solution from tube 3(ml)	-	-	-	1.0	-
Volume of solution from tube 4(ml)	-	-	-	-	1.0
Volume of distilled water (ml)	4.5	1.0	1.0	1.0	1.0
Total volume (ml)	5	2	2	2	2
Conc.TBZ (μg TBZ / 10 μL)	2	1	0.5	0.25	0.125

- Place 2ml of fresh egg in each well or 24 –well plate
- Add 10 μl of TBZ solution of varying concentration to corresponding wells. control well receive only water
- Incubate for 24-48 hour at 27 °C
- Add 2 drops of lugol's iodine to each well
- Count egg (dead & embryonated) and hatched larve from each well
- Calculate ED50 value for the egg hatch by log probit analysis

ED50 value more than 0.1μg TBZ/ml is suggestive of BZ resistance

B. Evaluation of the Efficacy and Resistance of Acaricides Against Ticks

Resistance in insecticide to a given acaricide or insecticide can be described as a reduction in susceptibility of a parasite to the acaricide when it is used at the recommended concentration and according to all of the recommendations for its.

The resistance of ticks to acaricides is an inherited phenomenon. It results from exposure of populations of ticks to chemical parasiticides (acaricides) and survival and reproduction of ticks that are less affected by the acaricide.

Detection of Acaricide Resistance in Ticks

In vitro test

1. Adult Immersion Test (AIT)

1. Weighing of engorged females of each replicate (4 replication with 5 females in each)

2. Preparation of different concentration of acaricides
3. Immersion of ticks for 2 minutes in different concentration of acaricides
4. Separate replication do negative control to be maintained
5. Dry the treated ticks on tissue paper and place 10 ticks/petridish
6. Treatment group for each concentration [(5 ticks×4 petridish) Negative control group (5 ticks×4 petridishes)]
7. Place the petri dishes in dessicator kept at 28^{0}C and 85 ± 5% RH (Separate dessicator for treated and control ticks)
8. Observe to ensure loss of pedal reflex followed by blackening, shrinkage and death in treated ticks (Observation to be taken up to 14 days)
9. Control tick show normal greenish appearance and commencement of oviposition
10. Calculation and data analysis

2. Larval Packet Test (LPT)

1. Prepare the packets of parallelogram shape (5.4 cm×5 cm)
2. Impregnate the packets with 500-600µl of desired concentration of acaricides and dry for 30 minute in incubator at 37^{0}C
3. Fold the rectangles in half and sealed on the sides with adhesive tapes forming an open-ended packet
4. Insertion of approximately 100-150 larvae (2 week old) sealed and placed in desiccators kept in BOD incubator maintained at 28 ± 1^{0}C and 85 ± 5% RH
5. The packets are open after 24 hrs, and larval mortality was calculated
6. Calculation and data analysis

C. Anticoccidial drug resistance

The World Health Organization Scientiûc Group (World Health Organization, 1965) developed the definition of resistance in broad terms as 'the ability of a parasite strain to survive and/or to multiply despite the administration and absorption of a drug given in doses equal to or higher than those usually recommended but within the limits of tolerance of the subjects'. Such a general definition could be accepted as a basis for the discussions on coccidiosis.

Diagnosis of Anticoccidial drug resistance

Diagnosis of resistance is generally performed by infecting groups of birds with suspect resistant *Eimeria* and comparing growth rates and pathology in these birds with those in birds infected with susceptible *Eimeria* in the presence of drug treatment. Resistance is reported as an anticoccidial index which is calculated using weight gain, % bird survival, lesion score and oocyst count. The lack of an *in vitro* assay for resistance to coccidia is a major impediment to research. The only accurate method available to obtain such information is the *in vivo* anticoccidial sensitivity test (AST).

The birds should be observed daily and deceased birds should be subjected to postmortem examination. At six days post-inoculation (D+6) necropsy should be performed on five birds per experimental group.

The diagnosis is mainly based on the methods like:

A) Coccidial lesion scores

B) Oocysts per gram of faeces (OPG)

C) Faecal/drapping score

E. tenella lesion score

- Score 1: Shows few scattered petechiae on the caecal wall
- Score 2: Noticeable blood can be found in the caecal contents, while the caecal wall is thickened
- Score 3: Large amounts of blood or caecal cores are present, caecal walls are greatly thickened
- Score 4: Caecal walls are enormously distended with blood or large caseous cores

E. brunetti lesion score

- Score 1: Stagnation of faecal contents with mild ballooning of intestine
- Score 2: Slightly greyish intestinal wall, thickening of lower intestine with fecks of salmon colour material sloughed off the intestine
- Score 3: Moderate thickening of intestinal wall with blood tinged catarrhal exudates; Transverse red, streaks markedly visible in lower rectum
- Score 4: Severe lesions extending up to the middle intestine

E. necatrix lesion score

- Score 1: Scattered petechiae and white spots observed on the serosa of intestine
- Score 2: Ballooning of mid gut portion
- Score 3: Pin point haemorrhages in the serosa with ballooning up to the lower part of the small intestine
- Score 4: Massive haemorrhages with brownish mucus contents; complete ballooning of the mid intestine

E. acervulina lesion score

- Score 1: Feature ladder-like white streaks in the duodenal loop ($5/cm_2$)
- Score 2: The lesion density is higher but not coalescent
- Score 3: The lesions are more numerous and coalesce, while thickening of the intestinal wall is visible
- Score 4: The mucosal wall is greyish with lesions completely coalesced

E. maxima lesion score

- Score 1: Characterised by few small red petechiae in the mid-intestine
- Score 2: More numerous petechiae are found and the intestinal content may be orange
- Score 3: The intestinal wall might be ballooned and thickened with pinpoint blood clots and mucous filling the intestinal contents
- Score 4: The intestinal wall is thickened and ballooned over most of its length, containing numerous blood clots and digested red blood cells

Thereafter, the mean lesion score per *Eimeria* spp. and experimental group should be calculated. In fresh-collected faeces material (from 4D to till D+8) the OPG-countings should be performed by means of a modification of the McMaster counting chamber technique.

Interpretation and results

The control should be based on the reduction of the mean lesion score of the medicated group compared to the infected untreated control (IUC) group using the formula: 100% - (mean lesion score of treated group/mean lesion score of the IUC group x 100%). A reduction percentage in the medicated birds compared with the IUC birds.

- 0-30% indicates coccidial resistance
- 31-49% reduction indicates reduced sensitivity or partial resistance
- 50% or more indicates full sensitivity to the tested anticoccidial drug

Further Readings

Bhatia, B.B., Pathak, K.M.L. and Juyal, P.D. (2010) Textbook of Veterinary Parasitology. 2nd Edition, Kalyani Publisher, Ludhiana.

Bowman, D.D. (2008) Georgis' Parasitology for Veterinarians. 9th Edition, Saunders, Elsiever.

Elsheikha, H.M. and Khan, N.A. (2011) Essentials of Veterinary Parasitology, Caister Academic Press.

Foreyt, W.J. (2001) Veterinary Parasitology: Reference Manual. 5th Edition, Blackwell Publishers, Ames, Iowa.

Hendrix, C.M. and Robinson, E. (2006) Diagnostic Parasitology for Veterinary Technicians. 3rd Edition, Mosby, Elsiever.

http://www.biogal.co.il/biogal-en/products/kits_for_companion_animals

Kaufmann, J. (1996) Parasitic infections of domestic animals - A Diagnostic Manual.Birkhauser Verlag AG, Basel, Switzerland.

Mullen, G. and Durden, L. (2009) Medical and Veterinary Entomology. 2nd Edition, Academic Press, Elsiever, London

Soulsby, E.J.L. (1982) Helminths, Arthropods and Protozoa of Domesticated Animals. 7th Edition, ELBS and Bailliere Tindall, London.

Taylor, M.A., Coop, R.L. and Wall, R.L. (2007) Veterinary Parasitology. 3rd Edition, Blackwell Publishers.

Urquhart, G.M., Armour J, Duncan JL, Dunn, AM and Jennings FW (2000) Veterinary Parasitology. 2nd Edition, Blackwell Publishers, Glasgow.

Zajac, A.M. and Conboy, G.A. (2012) Veterinary Clinical Parasitology. 8th Edition, Wiley Blackwell.

Appendix

Table 1: Characteristic features of important parasites/developmental stages

Group	Parasite	Salientfeatures
Helminthic eggs (Trematodes, cestodes and nematodes)	*Fasciola* spp.	Oval shape, golden yellow in colour, indistinct operculum and embryonic mass, size is 130 - 145µmx 70-90 µm
	Amphistomes	Oval shape, grayish-white colour, distinct operculum and embryonic mass, small knob is present on the end opposite to operculum, size is 160 µm x 90 µm
	Moniezia expansa	Some what triangular in shape, well-developed pyriform apparatus with hexacanth embryo
	Moniezia benedeni	Rectangular or quadrangular in shape, rest similar to *M. expansa*
	Dipylidium caninum	Eggs lie in egg capsule which is some what globular in shape, egg coat albuminous, inside which there is embryophore and oncosphere covering the hexacanth embryo
	Taenia spp.	Globular in shape, oncosphere has hexacanth embryo, embryophore has radial striations
	Ascaris suum	Oval or globular in outline, thick shelled, outer albuminous layer has prominent projections or corrugations
	Toxocara spp.	Sub-globular in outline, thick shelled and outer albuminous layer is finely pitted
	Oxyuris equi	Elongated in shape, slightly flattened on one side, plug is present at one pole
	Strongyloides spp.	Oval with blunt ends, thin shelled and contain a fully developed larva when laid
	Strongyles	Oval in shape, thin shelled and segmented embryo when laid
	Ancylostoma caninum	Thin shelled and contain about eight cell embryo when passed in faeces
	Trichuris spp.	Brown in colour and barrel shaped with a transparent plug at either pole
	Dioctophyma renale	Barrel shape, shells are pitted except of poles, unsegmented when laid
	Stephanurus dentatus	Ellipsoidal shaped, thin shelled, segmented when laid
	Capillaria plica	More barrel-shaped, sides nearly parallel, shell colourless, bipolar plugs do not project far
	Paragonimus westermanii	Yellowish-brown in colour, operculuted and the shell is thickened at the opposite pole
	Schistosoma nasale	Boomerang shaped and large in size
	Dirofilaria immitis (larvae)	Microfilariae (in blood) are sheathed, have long slender tail with a tapered anterior end and blunt posterior end
Protozoa	*Trypanosoma evansi*	Undulating membrane well developed and free flagellum is prominent, nucleus is centrally placed, kinetoplast is sub terminal and placed at posterior end

	Hepatozoon canis	In gametocytes, cytoplasm stains pale blue and surrounded by a delicate capsule around it, usually present in neutrophils
	Babesia bigemina	Typically the tapering end of the pyriform parasites jointed at an acute angle in erythrocyte
	Babesia equi	Forms maltese cross i.e. four parasites are present in a erythrocyte
	Babesia canis	Pointed at one end and rounded at the other end and lies at an acute angle, multiple infections of erythrocytes are common (upto16 parasites)
	Babesia gibsoni	The smaller form of canine *Babesia* having size 1.5-2.0μm and the parasites are generally oval , circular in shape
	Theileria annulata	Piroplasm is pleomorphic in shape viz. round, oval, rod like, comma shaped or dots present in erythrocytes; schizonts (Koch's blue body) present in the lymphocytes
	Tritrichomonas foetus	Organisms have pear shaped body, three anterior and a long trailing posterior flagellum present
	Eimeria spp.	Oocysts of different shape and size have 4 sporocysts with each having 2 sporozoites
	Isospora spp.	Oocysts have 2 sporocysts each with 4 sporozoites
	Cryptosporidium spp.	Sporulated oocyst with 4 sporozoites
Arthropods (Ticks and Mites)	*Boophilus microplus*	Short mouthparts, adanal plates and prominent caudal process present in male
	Hyalommaa. anatolicum	Longer mouth parts, first coxa is bifid, festoons and sub-anal plates present in males
	Riphicephalus sanguineus	Basis capituli is hexagonal, deep groove in first coxa, festoons present
	Demodex canis	Elongate and cigar shape, thorax has four pairs of stumpyl egs, abdomen with transverse striations
	Sarcoptes scabiei	Body circular, dorsal surface with fine folds and grooves with small triangular scales/ spines, legs are short and the posterior pair do not project beyond the margin of the body
	Psoroptes ovis	Oval in shape and larger than *Sarcoptes*, dorsal surface of body devoid of scales and spines, all the four pairs of legs project beyond the body margins
Rickettsia	*Anaplasma marginale*	Dot like parasites is present at the margins of the erythrocytes usually in pairs or some times singly without any cytoplasm
	Ehrlichia canis	Light blue colored morula present in monocytes and neutrophils

Table 2: Predilection site of parasites infecting domestic animals

Site of predilection	Parasite group	Cattle/Buffalo	Sheep/Goat	Equine	Dog/Cat	Pig
Digestive tract (Flukes)	Trematodes	*Paramphistomum epiclitum* *Cotylophoron cotylophorum* *Gastrothylax crumenifer* *Fischoederius elongates* *F. cobboldi* *Carmyerius gregarious* *Ogmocotyle indica*	*Paramphistomum cervi* *Cotylophoron cotylophorum* *Gastrothylax crumenifer* *Ogmocotyle indica* *Ceylonocotyle streptocoelium* *Calicophoron calicophorum*	*Gastrodiscus aegyptiacus* *Gastrodiscus secundus* *Pseudodiscus collinsi*	*Echinochasmus perfoliatus* *Heterophyes heterophyes* *Metagonimus yokogawai* *Alaria alata* *Alaria canis*	*Fasciolopsis buski* *Gastrodiscus aegyptiacus* *Gastrodiscoides hominis*
	Cestodes (Tapeworm)	*Moniezia expansa* *M. benedeni* *Avitellina centripunctata*	*Moniezia expansa* *Avitellina centripunctata*	*Anaplocephala magna* *Anaplocephala perfoliata*	*D. latum* *Spirometra mansoni* *Dipylidium caninum* *Echinococcus granulosus*	
		Thysanosoma actinoides *Thysaniezia giardi*	*Stilesia globipunctata* *Thysanosoma actinoides* *Thysaniezia giardi*	*Paranaplocephala mamillana*	*Taenia ovis* *T. hydatigena* *T. taeniaeformis* *T. serialis*	
	Nematodes (Roundworm)	*Toxocara vitulorum* *Gongylonema pulchrum* *Mecistocirrus*	*Gongylonema pulchrum* *Gaigeria pachyscelis*	*Parascaris equorum* *Habronema muscae* *H. majus*	*Toxascaris leonina* *Toxocara canis* *T. cati* *Strongyloides*	*Ascaris suum* *Physocephalus sexalatus* *Simondsia paradoxa*

Contd.

	digitatus		*Draschia*	*stercoralis*	*Gongylonema*
	Haemonchus	*Haemonchus*	*megastoma*	*S. cati*	*pulchrum*
	contortus	*contortus*	*Gongylonema*	*Ancylostoma*	*Gnathostoma*
	H. placei	*H. placei*	*pulchrum*	*caninum*	*hispidum*
	Trichostrongylus	*Mecistocirrus*	*Strongyloides*	*A.tubaeforme*	*Strongyloides*
	colubriformis	*digitatus*	*westeri*	*A.braziliense*	*westeri*
	Ostertagia ostertagi	*Marshallagia*	*Strongylus*	*A. duodenale*	*S. ransomi*
	Cooperia punctata	*marshalli*	*vulgaris*	*Uncinaria*	*Trichinella spiralis*
	Nematodirus spathiger	*Trichostrongylus*	*S. edentatus*	*stenocephala*	*Oesophagostomum*
	Agriostomum vryburgi	*colubriformis*	*S.equinus*	*Necator americanus*	*dentatum*
	Bunostomum	*Ostertagia ostertagi*	*S. asini*	*Trichuris vulpis*	*Necator americanus*
	phlebotomum	*Cooperia punctata*	*Triodontophorus* spp.	*Capillaria* spp.	*Globocephalus*
	Strongyloides	*Nematodirus spathiger*	*Cyathostomum* spp.	*Physaloptera canis*	*urosubulatus*
	papillosus	*Oesophagostomum*	*Oxyuris equi*	*P. felidis*	*Trichostrongylus*
	Trichuris ovis	*columbianum*		*Gnathostoma*	*colubriformis*
	Chabertia ovina	*Chabertia ovina*		*spinigerum*	*Hyostrongylus*
	Oesophagostomum	*Bunostomum*		*Spirocerca lupi*	*rubidus*
	radiatum	*trigonocephalum*			*Mecistocirrus*
	Paracooperia	*Agriostomum* spp.			*digitatus*
	nodulosa	*Trichuris ovis*			
Protozoa	*Giardia bovis*	*Tetratrichomonas ovis*	*Tritrichomonas equi*	*Giardia canis*	*Tetratrichomonas*
	Entamoeba bovis	*Giardia caprae*	*Trichomonas caballi*	*Giardia cati*	*buttreyi*
	Eimeria bovis	*Entamoeba ovis*	*Giardia equi*	*E. histolytica*	*Tritrichomonas*
	E. zuernii	*Eimeria ahsata*	*Entamoeba*	*Isospora canis*	*suis*
	E. bombayensis	*E. arloingi*	*equibuccalis*	*Sarcocystis cruzi*	*Enteromonas suis*
	E. bareillyi	*E. gilruthi*	*Eimeria leuckarti*	*Sarcocystis ovifelis*	*Entamoeba suis*
	Cryptosporidium bovis	*E. ovina*	*E. solipedum*	*Isospora felis*	*Eimeria debleicki*
		Cryptosporidium agni		*Besnoitia besnoiti*	*Eimeria suis*
				Toxoplasma gondii	*Isospora suis*
				Neospora caninum	*Balantidium coli*

Liver	Trematodes	*Dicrocoelium dendriticum* *Fasciola gigantica* *Fasciola hepatica* *Fascioloides magna* *Gigantocotyle explanatum*	*Dicrocoelium dendriticum* *Eurytrema pancreaticum* *Fasciola gigantica* *Fasciola hepatica* *Fascioloides magna*	*Dicrocoelium dendriticum* *Fasciola gigantica* *Fasciol haepatica* *Fascioloides magna*	*Dicrocoelium dendriticum* *Opisthorchis tenuicollis* *Clonorchis sinensis* *Fasciola hepatica* *Fasciolopsis buski*	*D. dendriticum* *Opisthorchis tenuicollis* *Clonorchis sinensis* *Fasciola hepatica* *Eruytrema pancreaticum*
	Cestodes	*Stilesia hepatica*	*Stilesia hepatica*	*Cysticercus tenuicollis*		*Cysticercus tenuicollis*
	(Protozoa)	*Thysanosoma actinioides* *Cysticercus tenuicollis* Hydatid cyst	*Thysanosoma actinioides* *Cysticercus tenuicollis* Hydatid cyst	Hydatid cyst	*Hepatozoon canis* *Leishmania donovani* *L. tropica*	Hydatid cyst
Blood-vascular system	Helminths	*Schistosoma* spp. *Ornithobilharzia* spp. *Onchocerca armillata*	*Schistosoma* spp. *Ornithobilharzia* spp.	*Schistosoma* spp. *Ornithobilharzia* spp. *Strongylus vulgaris* (larvae)	*Schistosoma* spp. *Ornithobilharzia* spp. *Dirofilaria immitis*	*Schistosoma* spp.
	Protozoa & Rickettsia	*Trypanosoma evansi* *T. theileri* *Babesia bigemina* *Theileria annulata* *T. buffeli* *Anaplasma marginale*	*Trypanosoma* spp. *Babesia* spp. *Theileria* spp.	*Babesia* (*Theileria*) *equi* *Trypanosoma evansi*	*Trypanosoma evansi* *Babesia canis* *B. gibsoni* *Ehrlichia canis* *Hepatozoon canis*	*Trypanosoma evansi* *T. simiae* *Babesia* spp.
Urogenital system	Helminths	-	-	*Dioctophyma renale* *Capillaria plica*	*Stephanurus dentatus*	
	Protozoa	*Tritrichomonas foetus*	-	*Trypanosoma equiperdum* *Klossiella equi*	-	-

Contd.

Respiratory system	Helminths	Hydatid cyst *Dictyocaulus viviparus* *Schistosoma nasalis*	Hydatid cyst *P. westermanii* *Dictyocaulus filaria* *Protostrongylus rufescens* *Muellerius capillaris* *Oestrus ovis* (fly larvae)	Hydatid cyst	*Paragonimus westermanii* *Linguatula serrata* (arthropod)	*Paragonimus westermanii* *Metastrongylus elongatus*
Skin and Sub-cutaneous tissue	Helminths	*Stephanofilaria assamensis* *S. kaeli* *S. zaheeri* *Parafilaria bovicola* *Onchocerca* spp.	-	*Habronema* spp. *Parafilaria multipapillosa*	*Dirofilaria* spp.	-
	Arthropods	*Hypoderma* spp. *Damalinia bovis* *Haematobia* spp. *Hippobosca* spp. *Linognathus* spp. *Haematopinus* spp. Ticks *Psoroptes* spp. *Sarcoptes scabiei* *Demodex bovis*	Blowflies larvae (maggot) *Damalinia* spp. *Linognathus* spp. *Haematopinus* spp. Ticks *Sarcoptes scabiei* *Psoroptes* spp. *Demodex ovis*	*Damalinia equi* Ticks *Sarcoptes scabiei* *Demodex equi*	*Heterodoxus spiniger* *Linognathus* spp. Fleas Ticks *Sarcoptes scabiei* *Demodex canis*	Ticks *Sarcoptes scabiei* *Demodex* spp. *Haematopinus* spp.
	Protozoa	*Besnoitia besnoiti*	-	*Besnoitia besnoiti*	-	-
Eye	Helminths	*Thelazia rhodesii*	*Thelazia rhodesii*	*T. lacrymalis* *Setaria equina*	-	-
Nervous system	Helminths & protozoa	*Toxoplasma gondii* *Neospora caninum*	*Coenurus cerebralis* *Toxoplasma gondii*	Hydatid cyst *Setaria digitata* *Sarcocystis neurona*	-	-
	Muscles	Sarcocysts	Sarcocysts	Sarcocysts	-	*Trichinella spiralis*

Table 3: Commonly used antiparasitic drugs

Group	Disease/Parasitosis	Antiparasiticdrugs
Helminths Trematodes	Fasciolosis	Triclabendazole @12 mg/kg b. wt. (cow); 10 mg/kg b.wt.(sheep & goat) orally; Oxyclozanide @10-15 mg/kg b.wt. (cow & buffalo); 15mg/kg b. wt. (sheep & goat) orally; Rafoxanide @ 7.5 mg/kg wt. (cow, buffalo, sheep & goat) orally
	Amphistomosis	Oxyclozanide @ 18.7mg/kg b.wt. (cow & buffalo) orally two doses two days apart; Niclosamide @ 50-100 mg/ kg b.wt. (cow, buffalo & sheep) orally
	Schistosomosis	Praziquantel @ 10-15mg/kg b. wt. (cow, buffalo, sheep & goat) orally
	Nasal schistosomosis	Praziquantel @ 25mg/kg b. wt., 2 treatments 3-5 weeks apart orally; Triclabendazole @ 20 mg/ kg b. wt. orally
	Equine cestodoses	Niclosamide @ 50-100 mg/kg b. wt. orally; Pyrantel emboate @ 38 mg/kg b.wt. orally; Mebendazole @ 10-15 mg/ kg b. wt. orally
	Ruminant cestodoses	Praziquantel @10-15 mg/kg b. wt. orally; Niclosamide @ 50-100mg/kg b. wt. orally; Mebendazole @ 10-15mg/ kg b. wt. orally
	Canine cestodoses	Praziquantel @5-7.5 mg/kg b.wt.orally; Niclosamide @75-100mg/kg b.wt.orally; Mebendazole @ 22 mg/kg b. wt.orally; Fenbendazole @ 50mg/kg b.wt. orally
	Ascariosis	Pyrantel pamoate @ 5-10mg/kg b.wt. (dog) orally; 6.6 mg/kg b.wt. (horse) orally; 250 mg total dose (buffalocalf) orally; Pyrantel emboate @19mg/kg b.wt. (horse) orally;14.4mg/kg b.wt.(dog) orally; Ivermectin @0.2mg/kg b.wt. (horse) parenterally; 0.3mg/kg b.wt. (pig) parenterally; Piperazine @ 200-300mg/kg b.wt. (cow, buffalo, horse, pig & dog) orally; Benzimidazoles @5-15mg/kg b.wt.orally
	Horse strongyle infections	Oxfendazole @10mg/kg b.wt.orally; Fenbendazole @ 30-60mg/kg b.wt. single dose or 7.5-10mg/ kg. b wt.for 5 days orally; Albendazole @5-7.5mg/kg b.wt.orally; Ivermectin @0.2 mg/kg b.wt. parenterally; Mebendazole @10-15mg/kg b.wt. orally; Tetramisole @15mg/kgb.wt. orally
	Ruminant strongle infections	Albendazole @5-7.5mg/kg b.wt.orally; Fenbendazole @5-7.5mg/kg b.wt.single oral dose; Mebednazole @10-15mg/kg b.wt. orally; Tetramisole @15mg/kg b.wt. orally; Ivermectin @0.2mg/kg b.wt. parenterally
	Canine hookworm infection	Albendazole @5-7.5 mg/kg b.wt.orally; Fenbendazole @ 5-7.5mg/kg b.wt. orally
	Canine filarial	Diethylcarbamazine 6.6mg/kg b.wt. parenterally for 3-4 days;

Contd.

	worm infection	Macrocyclic lactones are also considered to be effective
	Bovine filarial worm infection	Tetramisole @ 15mg/kg b.wt. orally; Levamisole @ 7 mg/kg b. wt. parenterally; Oxfendazole @ 5mg/ kg b.wt. orally
Arthropods	Lice infestation	Cypermethrin(10%) @1ml/L of water as body spray or dip; 5ml/L as back line spray; 20ml/L as poultry house spray; Cypermethrin (1%) shampoo @ 5ml/L as spray on pets
	Fly larval infestations	Ivermectin (1%) @ 0.2mg/kg b.wt. S/C; Doramectin (1%) @ 0.2 mg/kg b.wt. S/C; Closantel @10mg/kg b.wt. orally; In case of stomach bots of equines the eggs deposited hair can be sponged with lukewarm water containing any insecticide while in case of oxwarbles the larva on the back may be removed by injecting 1ml of 3% H_2O_2 solution
	Fly infestation	Closantel @10-12mg/kg b.wt. orally; Ivermectin (1%) @ 0.2 mg/kg b.wt. S/C; Doramectin (1%) @ 0.2 mg/kg b.wt. S/C; Ointments for wound dressing
	Tick infestation	Deltamethrin (1.25%) @ 2ml/L as body spray or dip; Flumethrin (1%) @1ml/10kg b.wt. as pour on along the backline; Amitraz (12.5%) @ 2-4ml/L as body spray; Amitraz (5%) @ 6ml/L as body spray; Ivermectin (1%) @ 0.2 mg/kg b. wt. S/C; Doramectin (1%) @ 0.2 mg/kg b. wt. S/C
	Sarcoptic mange	Dettamethrin @4-6ml/L as body spray used as 2-3 weekly applications; Amitraz @3-4ml/L as body spray used as 2-3 weekly applications; Ivermectin(1%) @ 0.2 mg/kg b.wt.S/C; Doramectin (1%) @ 0.2 mg/kg b. wt. S/C; Eprinomectin @ 0.5mg/kg b.wt. Topical Appendix 65
	Psoroptic mange	Dettamethrin @4-6ml/L as body spray used as 2-3weekly applications; Amitraz @3-4ml/L as body spray used as 2-3 weekly applications; Ivermectin (1%) @ 0.2mg/kg b.wt. S/C; Doramectin (1%) @ 0.2mg/kg b. wt. S/C; Eprinomectin @ 0.5mg/kg b.wt. topical
	Demodectic mange	Dettamethrin @ 4-6ml/L as body spray; Amitraz (12.5%) @4ml/L as body spray (5-8 applications are generally required); Amitraz (5%) @ 10ml/L as body spray (5-8 applications are generally required); Ivermectin oral @0.3-0.6mg/kg b.wt. daily for 1-2 months; Ivermectin (1%) @ 0.2mg/kg b.wt.S/C; Doramectin (1%) @ 0.2 mg/kg b. wt. S/C; Eprinomectin @ 0.5mg/kg b.wt. topical
Protozoa	Ruminant coccidiosis	Amprolium (cow & buffalo) @ 10 mg/kg b. wt./day for 5 days & 5 mg/kg b. wt./day as prophylactic orally; Sulphamethazine (cow & buffalo) @ 140mg/kg b.wt. for 3 days & 5mg/kg b.wt./day for 15 days as prophylactic orally; 140mg/kg b.wt. for 3 days (sheep & goat) orally; Sulpha quinoxaline (cow & buffalo) @13mg/kg b.wt. for 14-21days as prophylactic orally; 0.015% in water for 3-5 days (sheep & goat) orally; Sulphaguanidine @ 0.2% in feed or 3g/lamb/day for 20 days orally; Nitrofurazone (cow & buffalo) @15mg/kg b.wt./day for 7 days orally; 0.04% infeed or 0.0133% in water for 7 days (sheep & goat) orally

	Pet animal coccidiosis	Amproliumized water (prepared by adding 30ml of 9.6% amprolium in 3.8 L of water) and given to pups for 7-10 days orally; Sulphadimethoxine @ 55 mg/kg b.wt. for 21 days orally
	Babesiosis	Diminazene aceturate @ 3.5 mg/kg b.wt. S/C or I/M for 2 days (cow & buffalo); @ 5mg/kg b.wt. S/C or I/M for 2 days (horse); @ 3-5 mg/kg b.wt. S/C or I/M (dog & cat); Imidocarb @ 6mg/kg I/M (dog)
	Theileriosis	Oxytetracycline @15mg/kg b. wt. I/M 4-6 times a day; Oxytetracycline LA @ 20mg/kg b. wt. I/M for every 4 days; Buparvaquone @2.5 mg/kg b. wt. I/M
	Surra	Antrycide methyl sulphate (10%) @ 5mg/kg b.wt. S/C; Antrycide prosalt (3.5g in 15 ml water) @ 7.4mg/kg b.wt. S/C; Suramin (10%) @ 4g/45kg b.wt. in horses either singly or in 3 divided doses over a period of 3 weeks I/V; 0.3-0.5g/45kg b.wt. in dogs I/V; 0.5g/45kg b.wt. at 4-6 weeks interval in cattle; Diminazene aceturate @3.5 mg/kg b. wt. S/Cor I/M; Isometamidium (1%, 2% or 4%) @ 0.5-1.0 mg/kg b. wt. I/M
	Canine hepatozoonosis	Trimethoprim-sulfadiazine @15mg/kg, orally two times, Clindamycin @10mg/kg, orally three times, and Pyrimethamine @0.25mg/kg, orally once used in combination for 14 days.
	Balantidial dysentery/ diarrhoeae	Tetracycline and Metronidazole are effective
Rickettsia	Anaplasmosis	Tetracycline@6-10mg/kg b.wt. I/M for 3 days
	Ehrlichiosis	Oxytetracycline or Tetracycline @ 66mg/kg b.wt. divided into 2-3 doses for 14 days; Doxycycline @ 10mg/kg b. wt. oral for 2-3 weeks

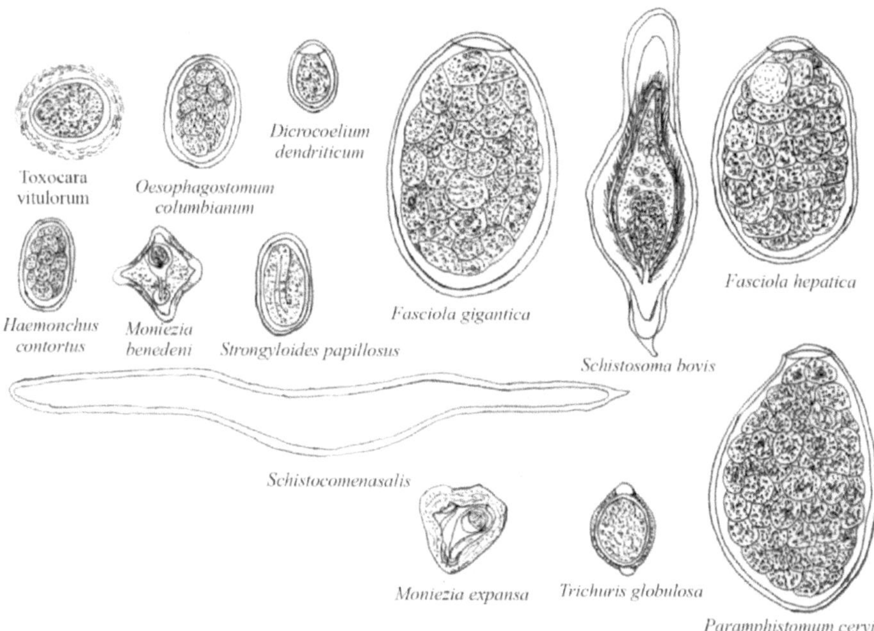

Eggs of commonly seen helminths of ruminants

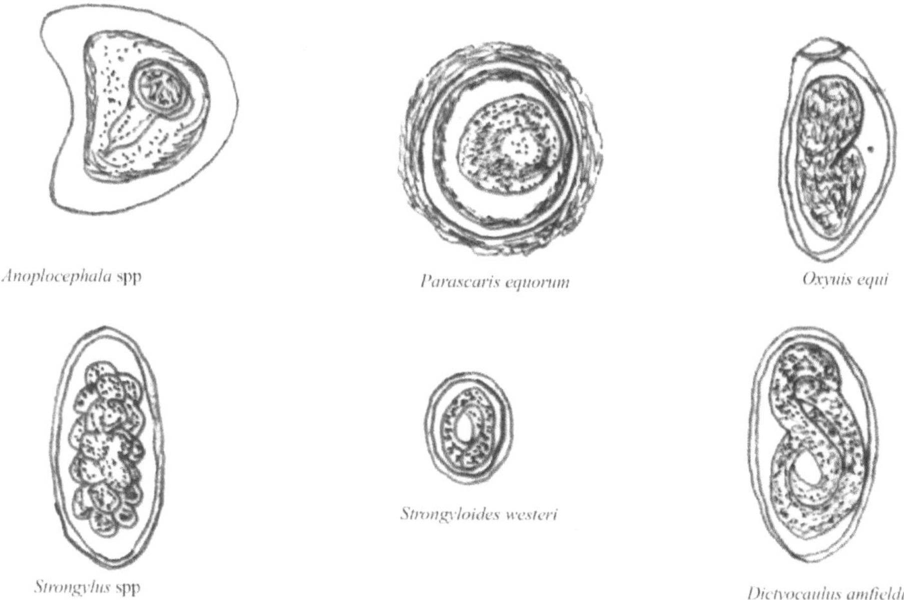

Eggs of commonly seen helminths of equines

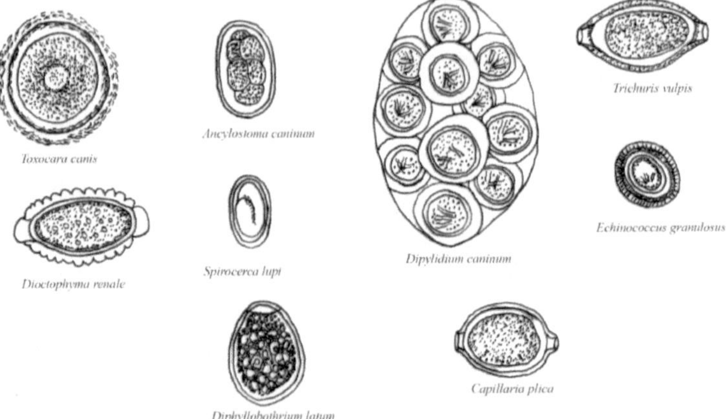

Eggs of commonly seen helminths of canines

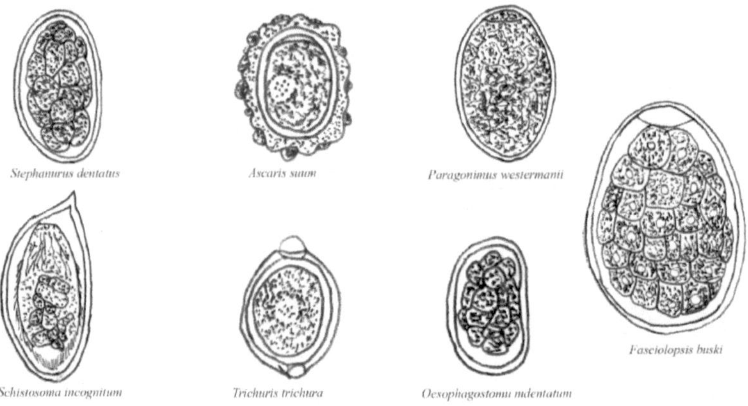

Eggs of commonly seen helminths of pigs

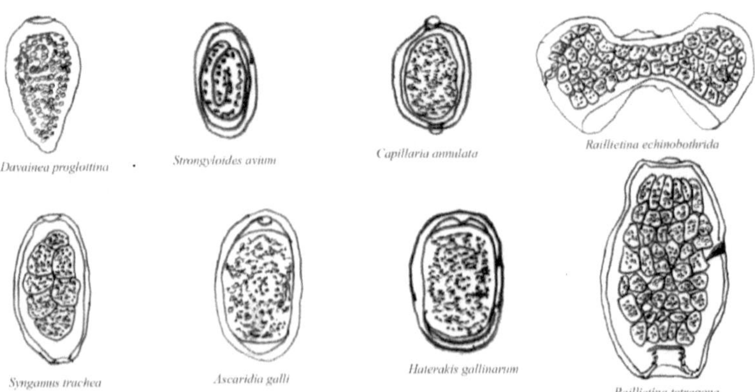

Eggs of commonly seen heminths of poultry

Commonly used chemicals/glasswares/ equipments in diagnostic laboratory

Equipments and glasswares Research microscope, Centrifuge machine Baermann's apparatus, McMaster chamber, Centrifuge tubes (15 ml)

Glass slides, Cover-slips,

Soft cloth (Muslin), Tissue paper,

Microscopic immersion oil, Plastic bottles,

Brushes and needles, Petri dishes, Specimen tubes, Pestle and mortar, Strainer,

Scalpel, Scissors, Forceps

Glass vials, Staining Rack

Chemicals

Absolute alcohol: To prepare 70% alcohol for preservation of parasitic specimen.

Formalin: Mix 10 ml of stock solution in 90 ml of water to prepare working solution.

Saturated salt solution: Mix the sodium chloride powder in water till saturation. The specific gravity of the salt solution is 1.2

Sheather's sugar solution

Cane sugar (ordinary)	:	560g
Thymol or phenol	:	6.5g crystal
Distilled water	:	320 ml
Specific gravity	:	1.25

This solution is highly efficient for coccidial oocysts, nematode ova and mange mites.

Zinc sulphate solution

Zinc sulphate	:	330g
Water	:	1000 ml
Specific gravity	:	1.18

Sodium nitrate solution

Sodium nitrate	:	400g
Water	:	1000 ml
Specific gravity	:	1.2

Magnesium sulphate solution

Magnesium sulphate	:	400 g
Water	:	1000 ml
Specific gravity	:	1.32

Giemsa stain (Stock solution)

Giemsa stain (powder)	:	750 mg
Glycerol	:	25 ml
Methanol (Pure)	:	75 ml

Transfer the powdered stain in a mortar, add glycerol, grind it with the pestle till a paste is formed, finally add methanol and mix thoroughly, transfer it to a dark coloured bottle and incubate the preparation at 37°C for 24 hours. Commercially prepared Giemsa stain solution is also available in the market that can be used as per the instructions of manufacturer.

Leishman stain (Stock solution)

Leishman stain (powder)	:	150 mg
Methanol (pure)	:	100 ml

Powdered stain and methanol are transferred to a dark coloured bottle and tightly closed with a glass stopper. The mixture is occasionally shaken gently for some days, before being used. Commercially prepared Leishman stain solution is also available in the market that can be used as per the instructions of manufacturer.

Wright's stain (Stock solution)

Wright's stain (powder)	:	0.3 g
Glycerol	:	3.0 ml
Methanol (Pure)	:	97 ml

Grind stain in a clean, dry mortar and then add glycerol until thoroughly mixed. Add methanol, keep for 24 hours and filter into clean bottle before use.

Anticoagulants: Following chemicals are used as anticoagulants

Trisodium citrate : 3.8 g

Distilled water : 100 ml

1 ml of this solution is used for 6 ml of blood.

Ammonium oxalate : 3 g

Potassium oxalate : 2 g

Distilled water : 50 ml

0.5 ml of the above mixture is needed for 10 ml of the blood.

Heparin : 5 I.U. / ml of blood

EDTA : 1 mg/ml of blood (ethylene diamine tetra-acetic acid)

Potassium dichromate solution: 2.5% solution used for sporulation of oocysts

Lugol's iodine solution

Potassium iodide : 10 g

Powdered iodine : 5 g

Distilled water : 100 ml

Filter and store in a tight stoppered bottle.

Diamond's medium (modified) for *in-vitro* culture of *Tritrichomonas foetus*

Trypticase peptone : 2 g

Yeast extract : 1 g

Maltose : 0.5 g

L-Cysteine hydrochloride : 0.1 g

L-Ascorbic acid : 0.02 g

Make up with 90 ml distilled water containing 0.08g each of K_2HPO_4 and KH_2PO_4, and adjust to pH 7.2-7.4 with NaOH or HCl. Following the addition of 0.05g agar, the media is autoclaved for 10 min at 121°C, allowed to cool to 49°C, and then 10ml inactivated bovine serum (inactivated by heating to 56°C for 30 minutes), 100,000 units crystalline penicillin G and 0.1 g streptomycin sulphate are added aseptically. The medium is aseptically dispensed in 10 ml aliquots into sterile vials and refrigerated at 4°C until use.

Hoyer's medium

Gum arabic	:	30 g
Glycerol	:	16 ml
Chloral hydrate	:	200 g
Distilled water	:	50 ml

Colour Plates

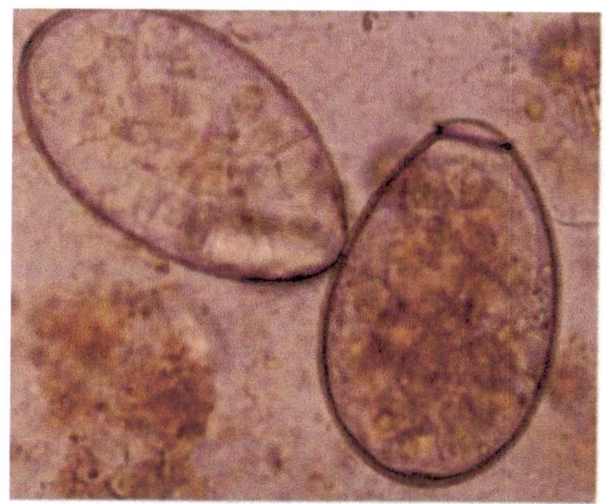

Eggs of Amphistome

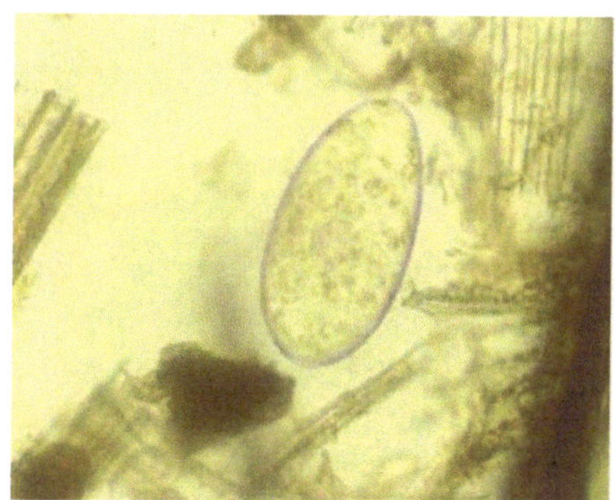

Egg of *Fasciola* spp.

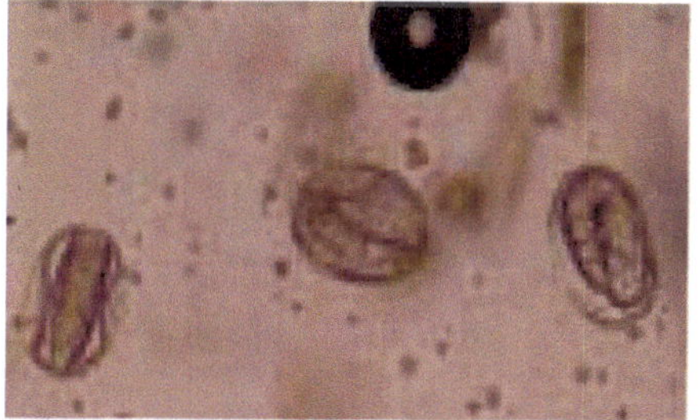

Eggs of *Strongyloides* spp.

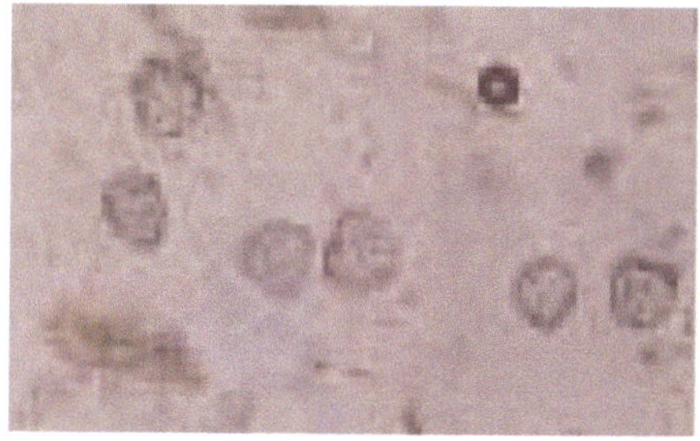

Eggs of *Moniezia expansa*

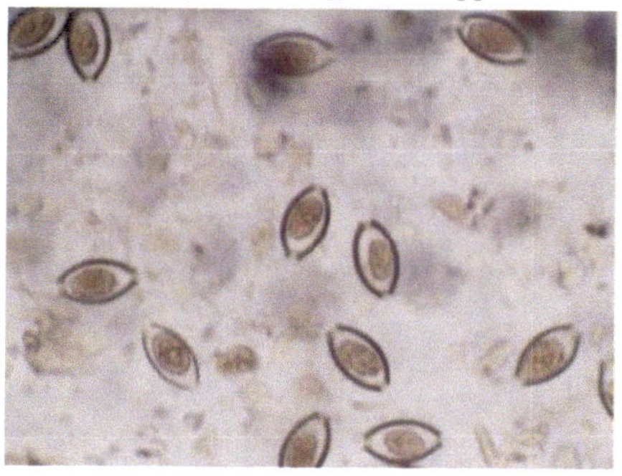

Eggs of *Trichuris* spp.

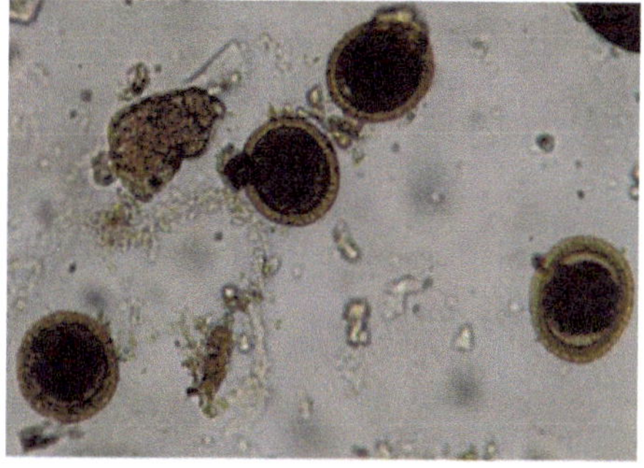

Eggs of *Toxocara vitulorum*

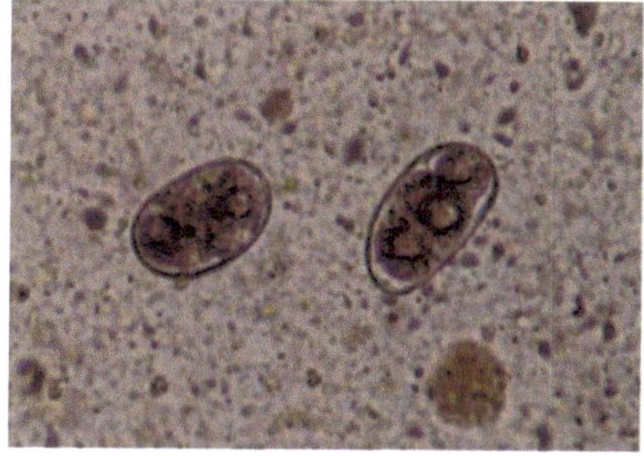

Eggs of Strongyles

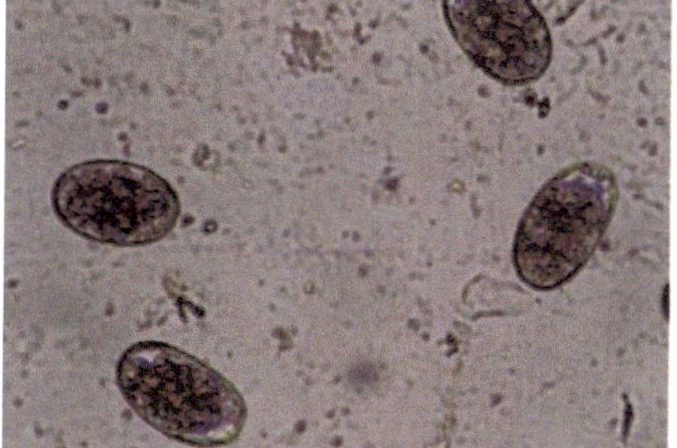

Eggs of *Ancylostoma* spp.

Plate I: Photomicrographs of various stages of importan thelminthic parasites (Cont).

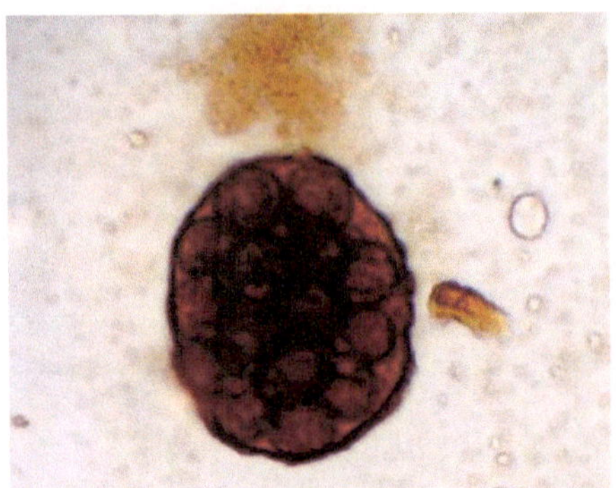

Egg packet of *Dipylidium caninum*

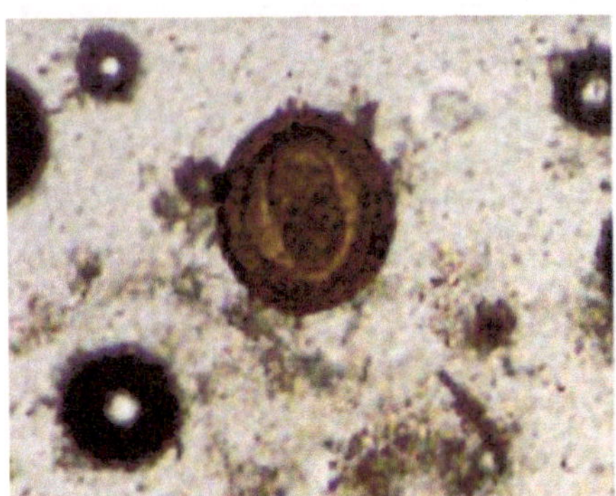

Egg of *Parascaris equorum*

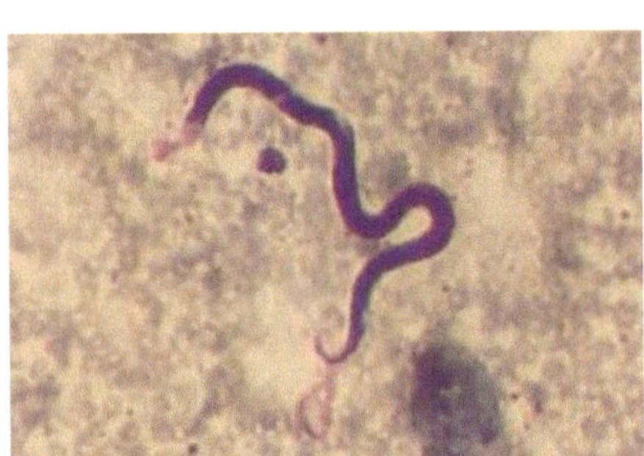

Microfilaria in blood of cattle

Adults of *Toxocara vitulorum*

Plate I: Photomicrographs of various stages of important helminthic parasites (Cont.).

Adults of *Parascaris equorum*

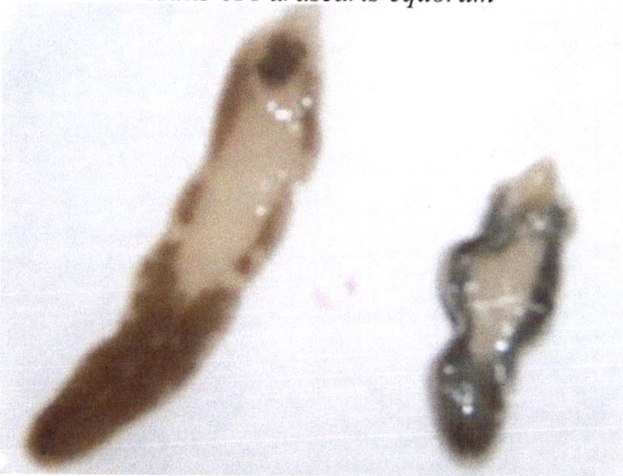

Adults of *Fasciola* spp.

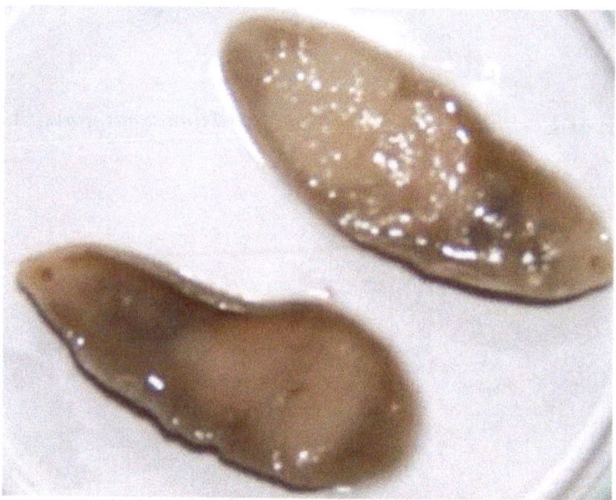

Adults of *Fasciolopsis buski*

Plate I: Photomicrographs of various stages of important helminthic parasites

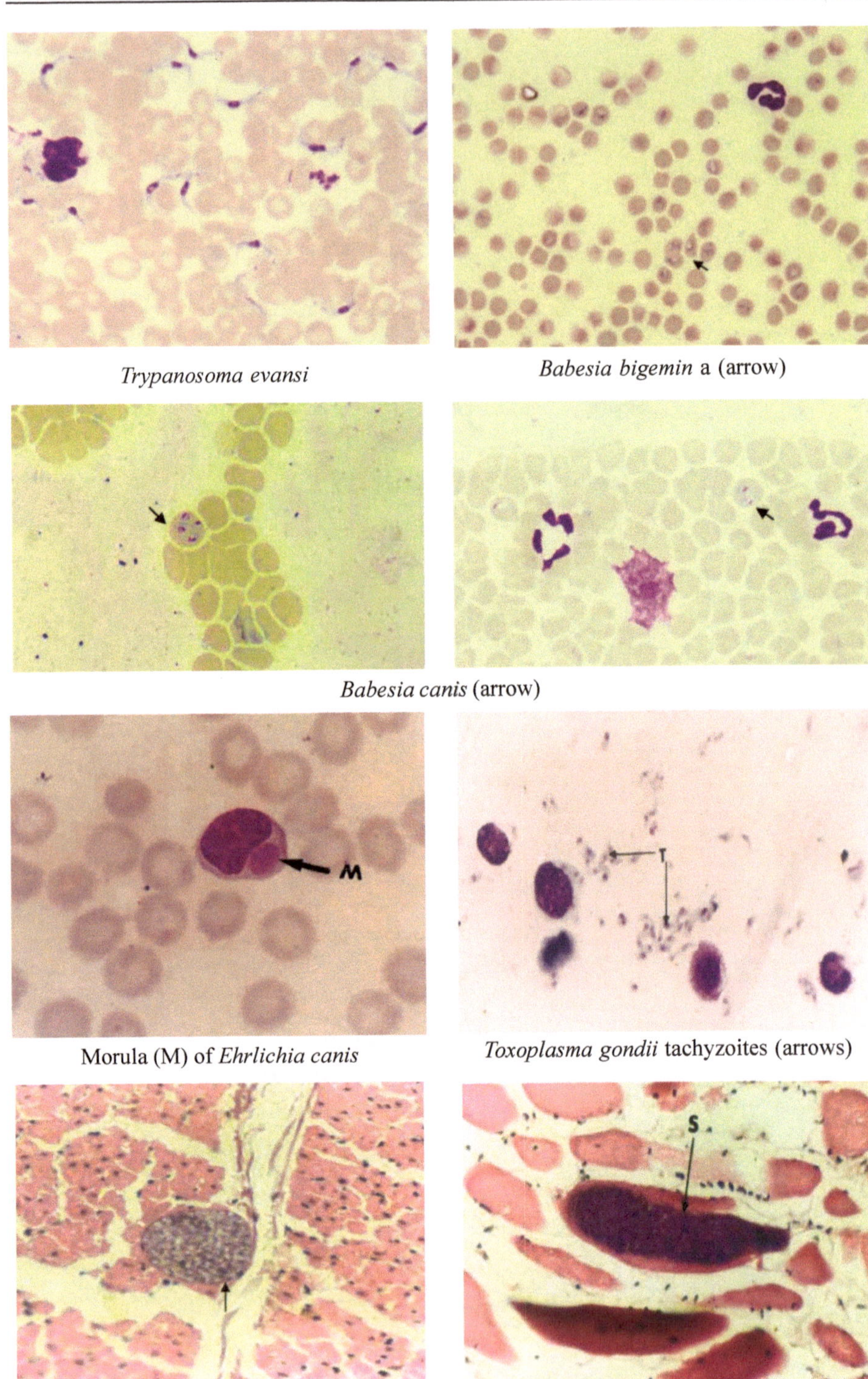

Trypanosoma evansi

Babesia bigemin a (arrow)

Babesia canis (arrow)

Morula (M) of *Ehrlichia canis*

Toxoplasma gondii tachyzoites (arrows)

Toxoplasma gondii cyst (arrow)

Sarcocystis cyst (arrow)

Plate II: Photomicrographs of various stages of important protozoa/Rickettsial parasites. (Cont.)

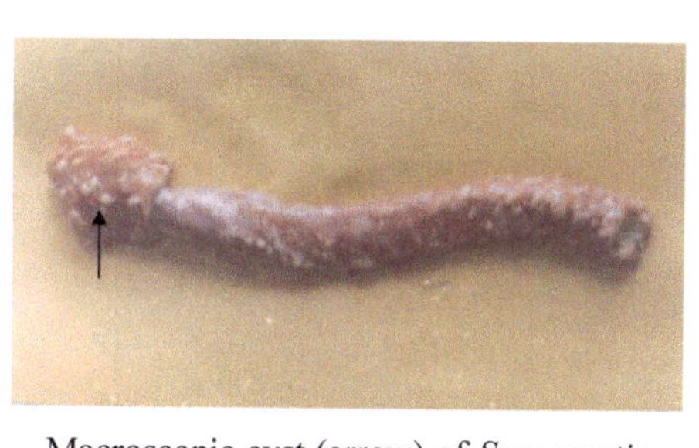

Macroscopic cyst (arrow) of *Sarcocystis fusiformis* in oesophagus of buffalo

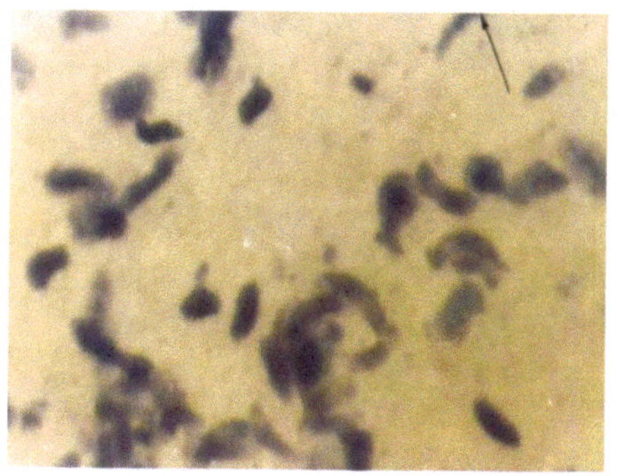

Bradyzoites of *Sarcocystis* (arrow)

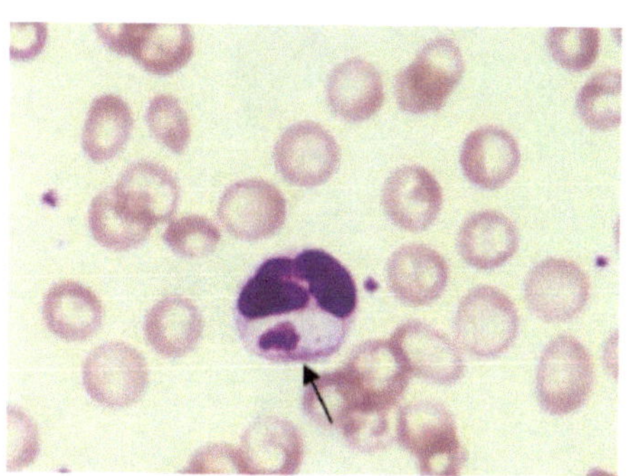

Hepatozoon canis (arrow)

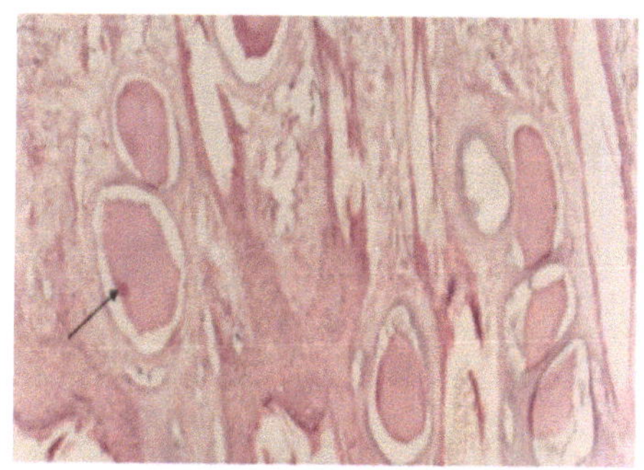

Besnoitia cyst (arrow)

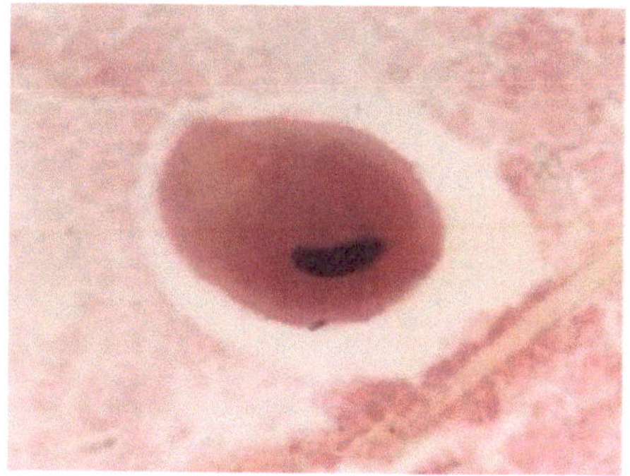

Balantidium coli (Cyst)

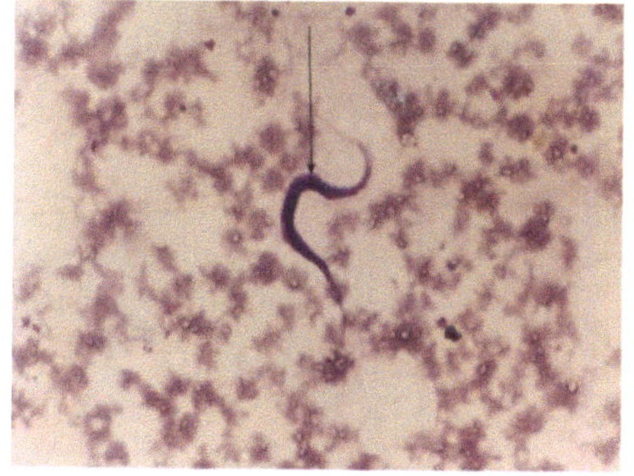

Trypanosoma theileri (arrow) in cattle

Plate II: Photomicrographs of various stages of important protozoa/Rickettsial parasites. (Cont.)

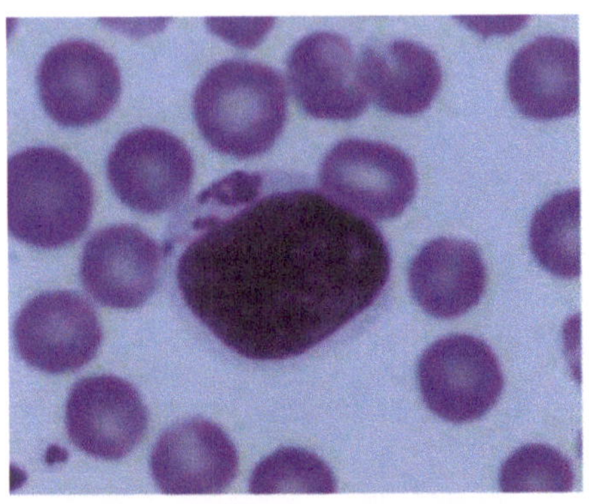

Koch's Blue Body (Schizont) of *Theileria annulata*

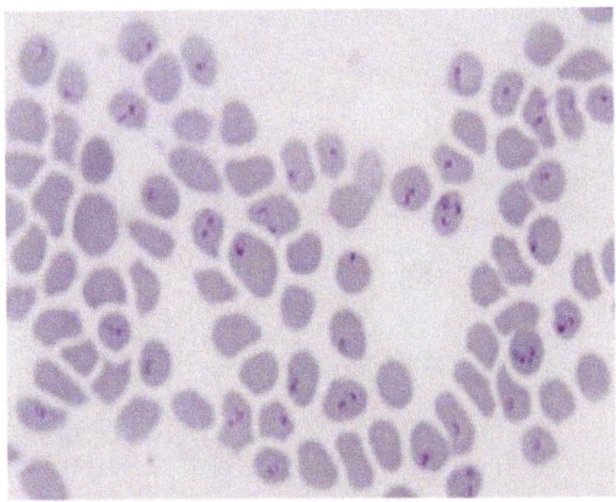

Theileria annulata piroplasms

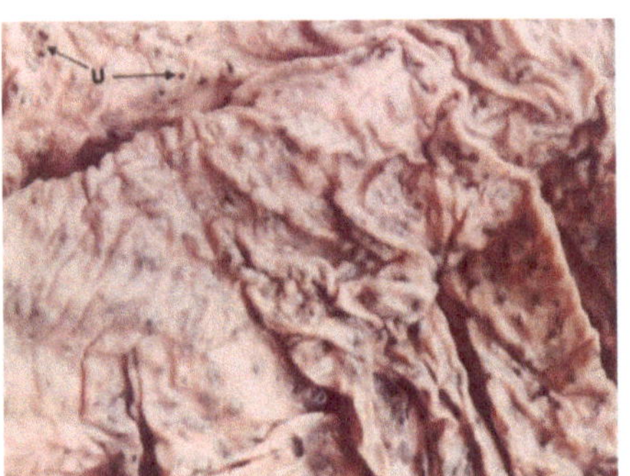

Abomasal ulcers (U) in *Theileria annulata* infecated calf

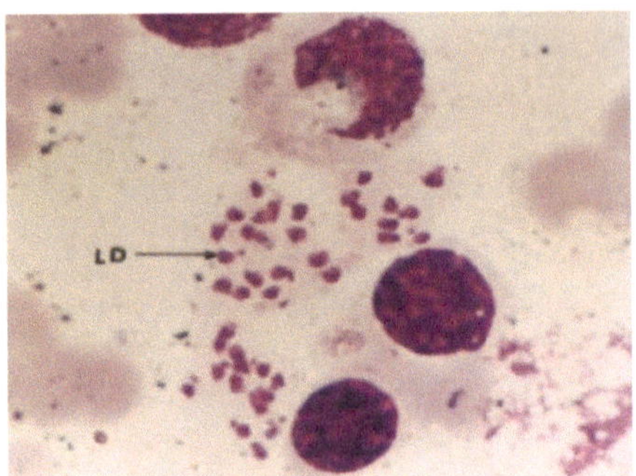

Leishman donovon (LD) bodies

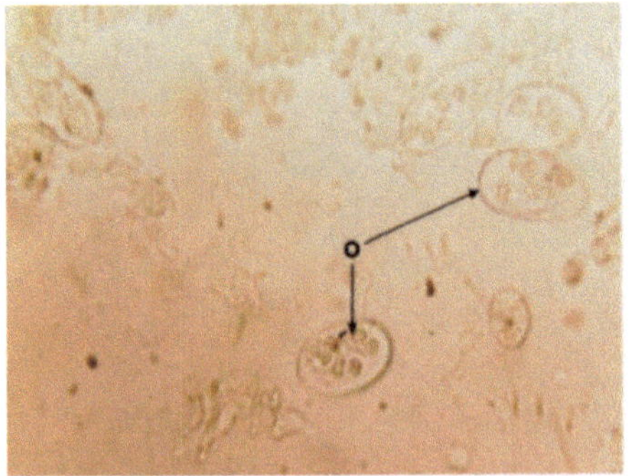

Sporulated coccidian oocyst (arrows)

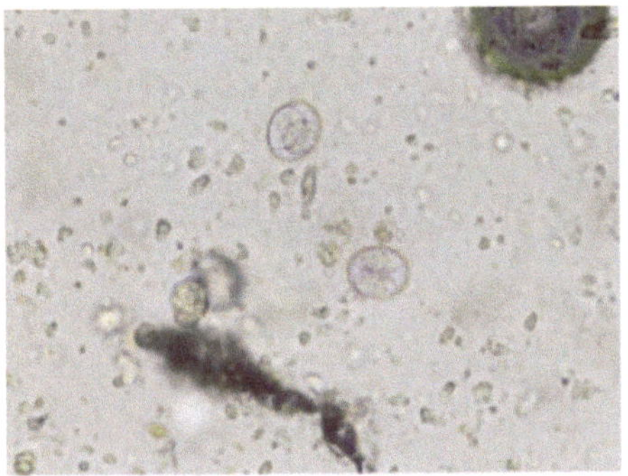

Isosporan oocysts (Sporulated)

Plate II: Photomicrographs of various stages of important protozoa/Rickettsial parasites. (Cont.)

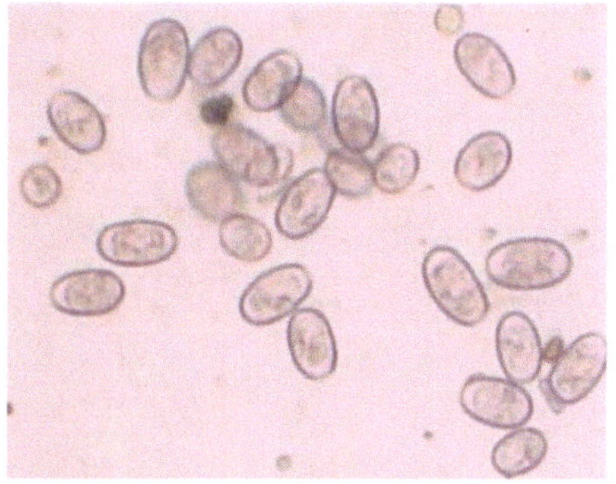

Coccidian oocysts (unsporulated)

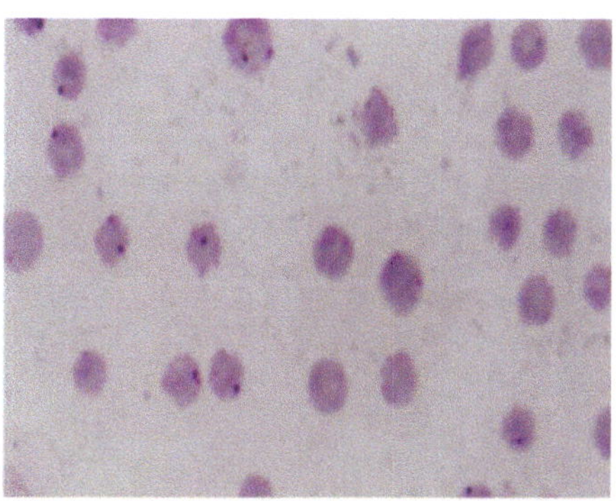

Anaplasma marginale

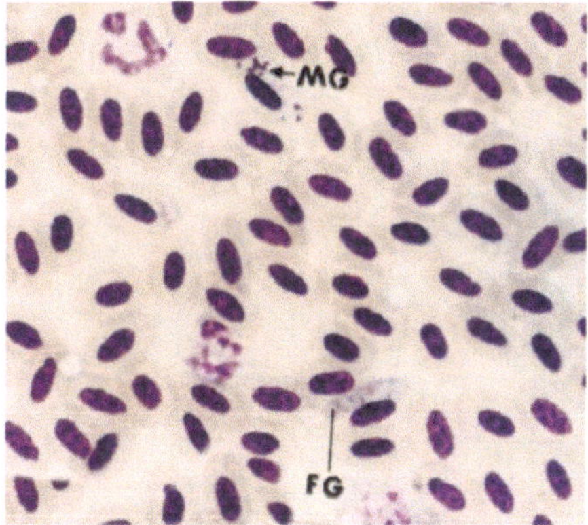

Male (MG) and Female (FG) gamont of *Haemoproteus columbae*

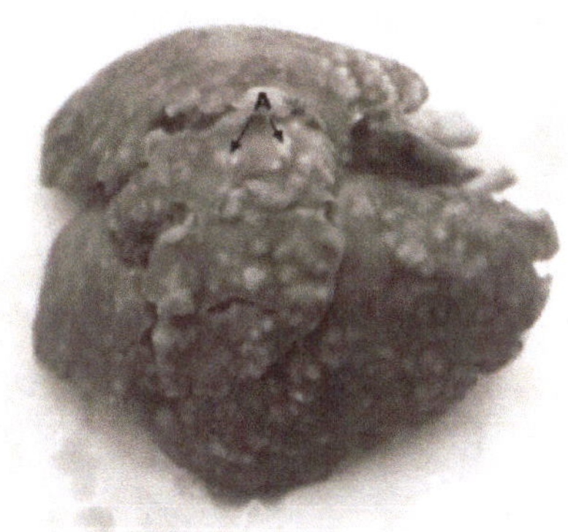

Abscesses (A) in liver in rabbit coccidiosis

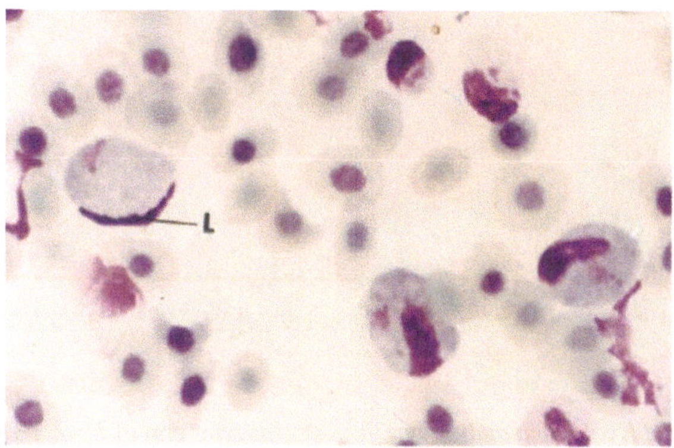

Leucocytozoon simondi (gamont L)

Plate II: Photomicrographs of various stages of important protozoa/Rickettsial parasites

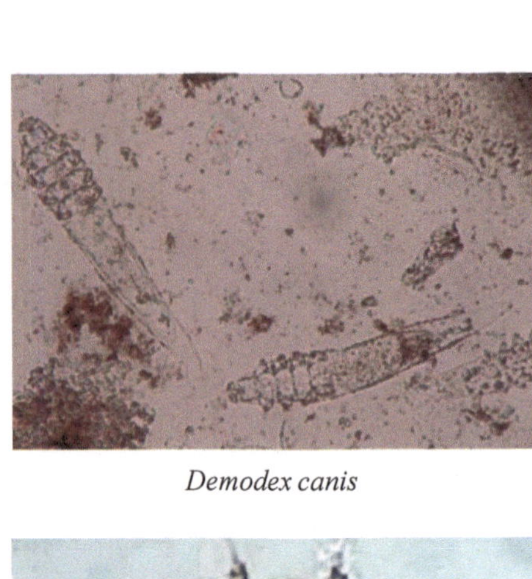

Demodex canis

Psoroptes spp.

Sarcoptes sp.

Adult male *Rhipicephalus microplus* showing prominent caual process (arrow)

Engorged female *Hyalomma anatolicum anatolicum*

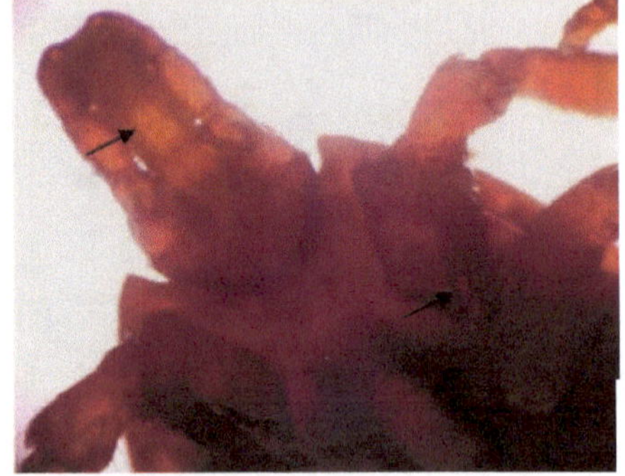

Anterior end of *H. a. anatolicum* showing longirostrate parts and Coxa-1bifid (arrows)

Plate III: Photomicrographs of various stages of important arthopod parasites (Cont.)

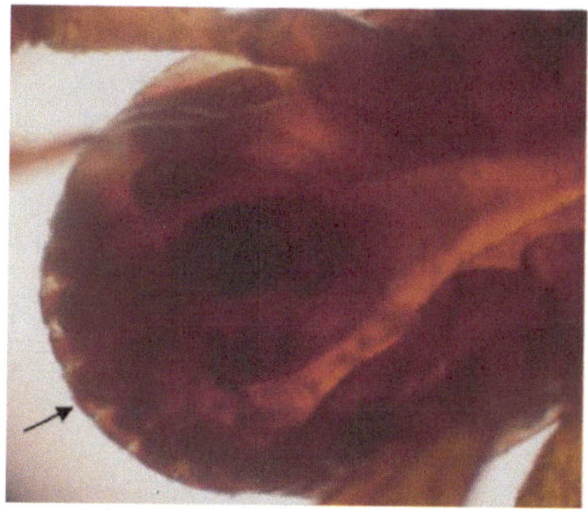

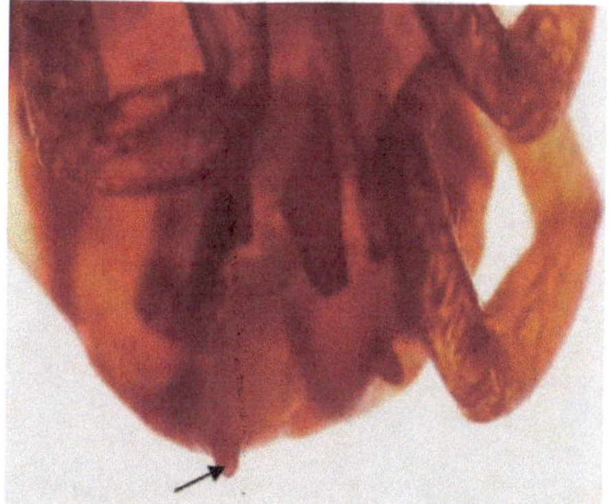

Posterior end of male *H. a. anatolicum* showing festoons and prominent sub-anal plates (arrow)

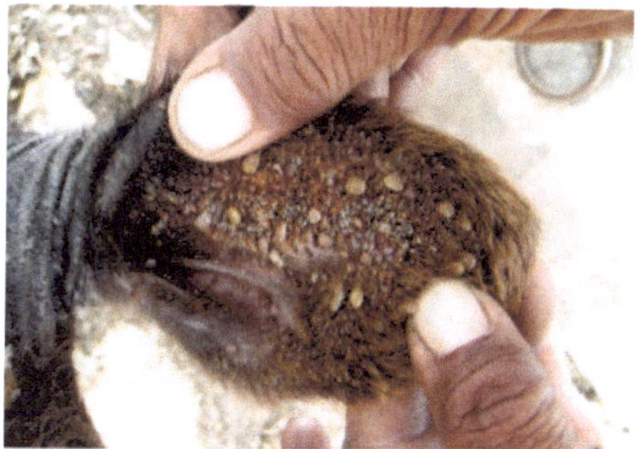

Pinna of cattle heavily infested with different stages of *Boophilus microplus*

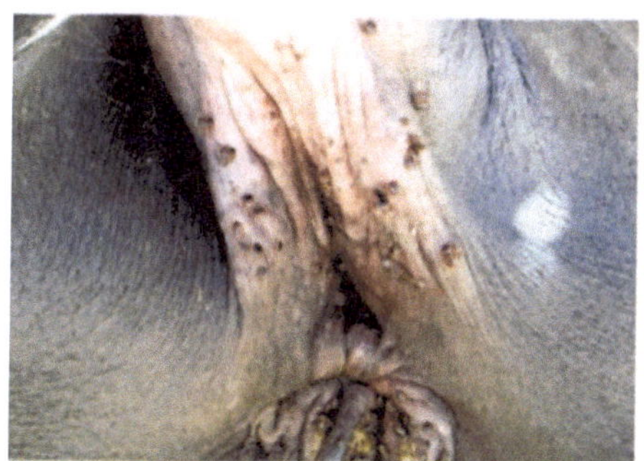

Base of tail of buffalo infested with *Hyalomma a. anatolicum*

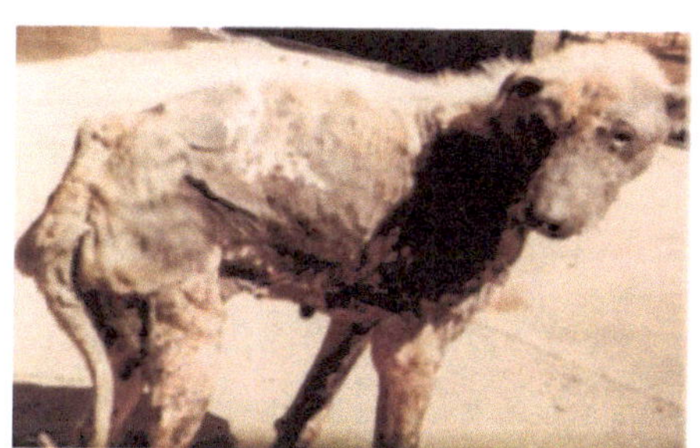

Severe lesions of demodectic mange in infested dog

Damalinia spp. (Lice)

Plate III: Photomicrographs of various stages of important arthopod parasites (Cont.)

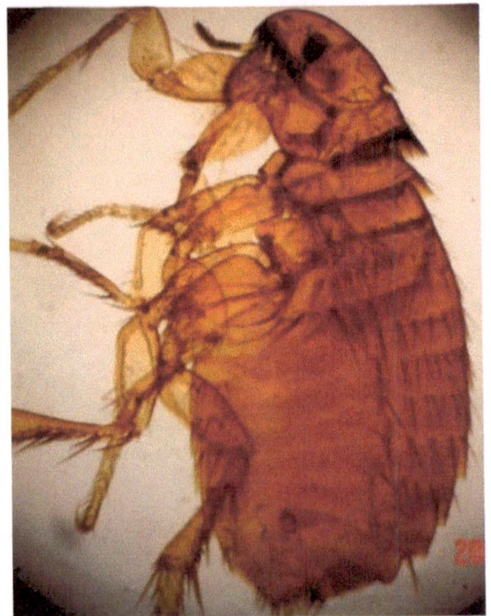

Ctenocephalides spp. (Flea)

Tabanus spp. (Horse Fly)

Sarcophaga spp. (Flesh fly)

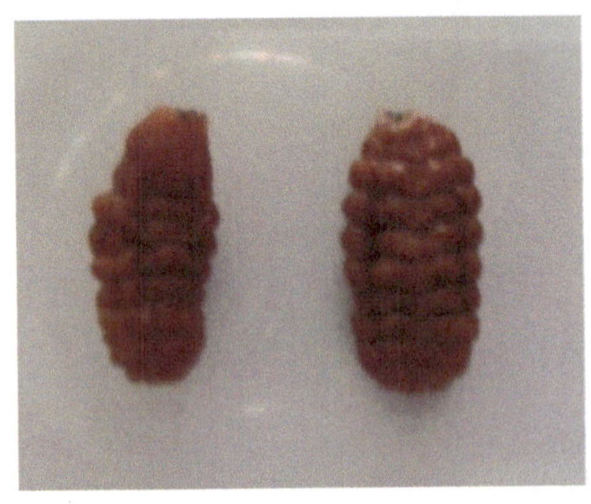

Hypoderma bovis larvae

Oestrus ovis larvae

Gasterophilus spp. larvae

Plate III: Photomicrographs of various stages of important arthopod parasites

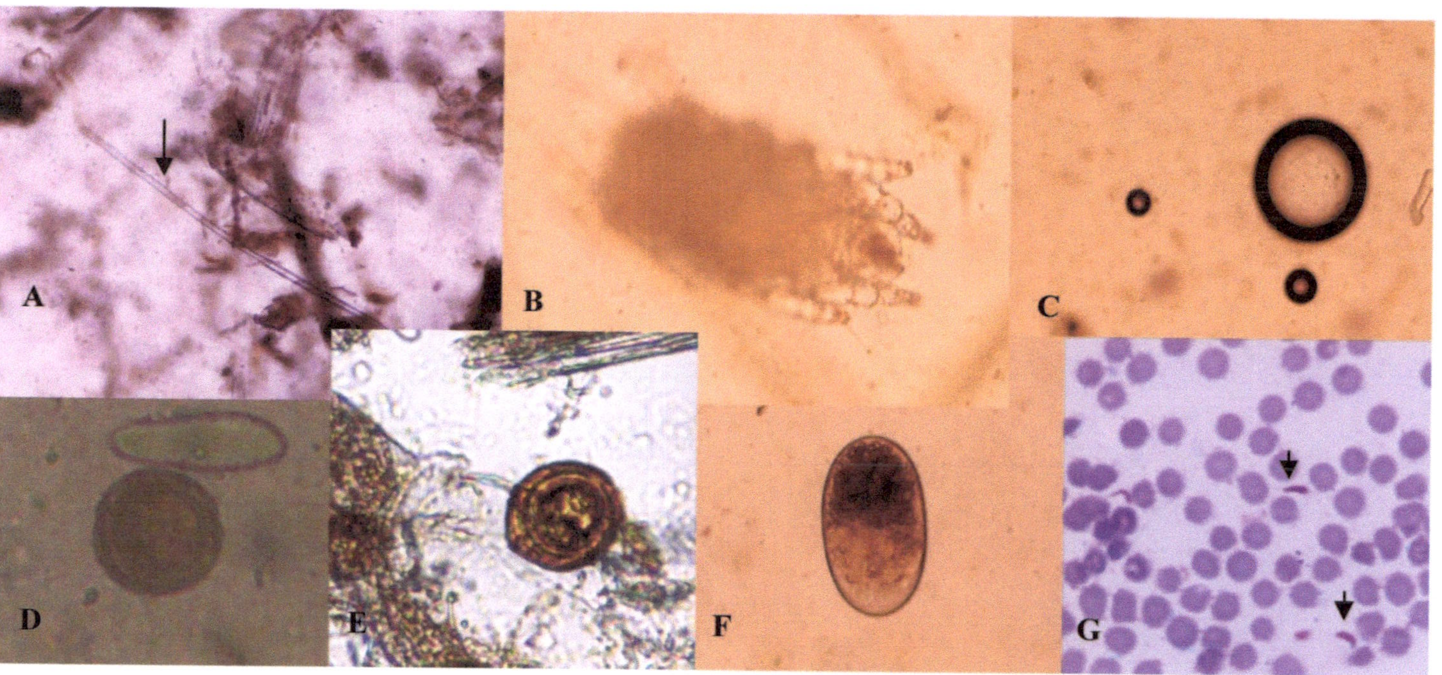

Plate IV: A: Plant hair (arrow, confused with nematode larva); **B:** Grain mite; **C:** Air-bubbles (confused; with eggs); **D:** Grain particle (confused with *Taenia* spp. egg); **E:** Plant pollen; **F:** Egg of grain mite (confused with Strongyle egg); G: Flagellated platelets (arrows, confused with *Trypanosoma evansi*)

Index

D

E

F

G

O

P

Q

R

S

T